Turtles, Termites, and Traffic Jams

Complex Adaptive Systems
John H. Holland, Christopher Langton, and Stewart W. Wilson, advisors

Turtles, Termites, and Traffic Jams

Explorations in Massively Parallel Microworlds

Mitchel Resnick

A Bradford Book
The MIT Press
Cambridge, Massachusetts
London, England

This book was set in Bembo by The MIT Press and was printed and bound in the United States of America.

Library of Congress Cataloging-in-Publication Data

Resnick, Mitchel.
   Turtles, termites, and traffic jams: explorations in massively parallel microworlds / Mitchel Resnick.
      p.   cm.—(Complex adaptive systems)
   Includes bibliographical references.
   ISBN 0-262-18162-2
   1. Parallel processing (Electronic computers) 2. Artificial intelligence.
3. StarLogo (Computer program language) I. Title. II. Series.
QA76.58.R47   1991
003'.7—dc20   94-10956
                                                                        CIP

In memory of my father . . .

Contents

Foreword

Seymour Papert

Resnick as Object-to-Think-With

No foreword could help make Mitchel Resnick's always lucid writing easier to follow. So instead of attempting the impossible, I'll try the opposite tack of making it more difficult—by inviting you to read more into the work than its author's modesty or intellectual caution allows him to claim explicitly. Stated bluntly, my thesis is that Resnick's work provides a rare glimpse of what I am sure will become a new paradigm for research in education.

There are many theories of what would be needed for computer-based media technologies to fulfill the optimistic early expectations that they would induce deep change in education. Some argue that the technology still needs time to mature; others say that we have enough technology, but research on media and education science needs more time and money; yet others place the emphasis on institutional and cultural resistance to change. Each of these theories has a large element of truth and will even find some explicit support in this book. But the existence of the book—taken as "an object to think with" rather than as an explicit argument—suggests a very different kind of deficiency. The critical bottleneck is not a shortage of machines or ideas but rather the shortage of people like Resnick who are able to innovate fluently in an unprecedented combination of media and willing to take the risks inherent to doing so.

I find it useful to situate Resnick's work in relation to two complementary fallacies that have contributed significantly to limiting thinking about the role of technology in education. The "technocentric fallacy" is illustrated by questions like: "How do computers change the way children learn?" Such questions must be rebutted rather than answered, for example by noting that computers don't change anything—people do—and people can use computers to quite different effects just as writing is used to very different effects in pornography and in romantic poetry. But a too-easy rebuttal leads into a complementary fallacy: the "just-a-tool fallacy." By this I mean the failure to distinguish between tools (reasonably described as "just tools") that improve their users' ability to do pre-existing jobs, and another kind of "tool" (of which this book offers an excellent example) that are more than "just tools" because of their role in the creation of a job nobody thought to do, or nobody could have done, before.

This distinction between kinds of tools cuts across the usual classifications of uses of computers in education. In some ways, "drill and kill" programs (like Math Blaster) are very different from "mathematics exploration programs" (like Geometric Supposer). One emphasizes rote memorization, the other mathematical investigation. But both can be considered as tools for well-established educational purposes. In both cases, the purpose can be clearly formulated in terms of established educational objectives. Tools of the other kind, the ones that do not serve a familiar established function, obviously pose a more difficult problem of description and of justification.

The title of this book suggests that StarLogo, the variant of Logo that Resnick has developed, is a tool for the study of turtles, termites, and traffic jams. In a sense it is. But educators who would immediately recognize the point of a tool for writing or for supporting "the math curriculum" might well wonder if they need a tool for this unfamiliar purpose. And they would probably still not immediately acknowledge their need even when it became clear that the title is a catchy figure of speech standing in for "liberating students from the confines of the centralized mindset." The point is that Resnick has identified an unacknowledged educational need as well as providing a tool to address it.

In this double achievement he anticipates a stage of development in the relationship between technology and learning for which education at large is only now reaching the threshold of readiness. All new technologies go through the same pattern of development. Their first uses are to do in a new way what was already being done; for example, the earliest movies were "photographed theater," and many of today's ideas about using computers in education simply see school's old methods through a computerized lens. It takes time for a new culture to emerge with new categories of people—in the case of cinema, the great directors, stars, and special-effects wizards—performing functions that were unimagined and largely unimaginable. The richest route to anticipating the analogous development in education is to look closely at the emerging new educators, and a good beginning is to focus on what Resnick needed to know to do what he did.

A prerequisite domain of knowledge for his work is his fluent mastery of the art of molding computer systems as media not only for expressing new ideas but even for cultivating them. Many people have used fluency in Logo to express ideas. It takes another level of computational fluency to change Logo in order to express what one has in mind; and even this does not capture what Resnick achieved, for although his project of creating StarLogo began with an idea about what he wanted to do with it, this idea was still far short of the rich web of ideas he evokes in his discussion of the "decentralized mindset." His thinking emerged, indeed in a way that illustrates his central thesis, as he worked in a field of varied intellectual influences. Among these some were technological: he drew on contemporary thinking about parallel and distributed computation and on access to the first really massively parallel computer. Others were scien - tific: his growing interest in decentralized processes had drawn him into a community of like-minded researchers on Artificial Life. Yet others were epistemological: beyond his involvement with theories of decentralized process, he has been sensitive to the growing contemporary culture of alternative ways of knowing. And finally I see the role of his educational interest as far more than the domain of application into which he channels ideas drawn from all these other sources. The chief lesson to be learned from Piaget is that thinking about how children think about X will, for almost all values of X, provide insights into mature thinking about X. Resnick's work provides a new and particularly elegant context for thinking about this two-way street of mutual enrichment between "education" and "science."

On a more personal level I would translate the two-way street model into the following advice. If reading Resnick's discussion of how children think about turtles, termites, and traffic jams doesn't tell you something new about your own thinking—try reading it again, this time with more care and openness.

Preface

When I was growing up, in a suburb of Philadelphia, there was a small field on the side of our house. On summer evenings, I would go to the "side lot" (as we called it), lie on my back, and stare into the sky. My eyes would dance from star to star. But it wasn't so much the stars that held my attention. Rather, it was the space between, around, and beyond the stars. At an early age (maybe seven or eight), I had started to wonder about all that space. Does it go on forever? If not, where does it end? How does it end?

Every answer that I could think of seemed equally absurd. I could not imagine the universe going on forever. But how could it end? If there is a wall at the end of the universe, what is on the other side? These questions frustrated and fascinated me. Of course, I came across many other questions that I couldn't answer. But for most questions, even if I didn't know the answer, I could at least imagine that there *was* an answer. Questions about the "end of the universe" took on a different status for me. I couldn't even imagine any answers. No answers seemed possible.

As I grew older, I became interested in puzzles and paradoxes. I spent many hours trying to sort out the sentence *This sentence is false.* (If the sentence is true, then it must be false! But if it is false, it must be true!) After a while, my mind would always wander back to my Ultimate Paradox, the paradox of a universe that can't go on forever but can never end.

In school, I was attracted to math and physics, two fields filled with paradoxes and counterintuitive ideas. As an undergraduate at Princeton University, I decided to major in physics, specializing in astrophysics and cosmology. I was determined to develop a better understanding of my Ultimate Paradox. In physics courses, I learned how to derive and manipulate the equations of general relativity, the field most directly related to my Ultimate Paradox. But it wasn't the equations that really interested me. The equations didn't provide the "answer" to the Ultimate Paradox. The equations were just a foundation, a jumping-off point, for thinking about it. I kept trying to develop new intuitions and new metaphors for thinking about the Ultimate Paradox. I learned that the universe might curve back on itself, just as the land on Earth curves back on itself as you travel all the way around the globe. But what does that mean? How can three-dimensional space "curve back on itself"? How could I envision that? How could I "feel" that?

During my senior year, I applied to graduate school in physics. But at the end of the year, I decided not to attend. I worried that physics graduate school would be filled with too many equations, and too few qualitative insights. Instead, I started working as a journalist, specializing in science and technology. I was still fascinated with the mysteries and paradoxes of science, and I wanted to share my fascination with others.

I enjoyed working as a journalist. I spent two years writing about universities and high-technology companies around Boston, then another three years writing about Silicon Valley. I learned about science and technology from a totally new perspective—and, through my writing, I tried to help other people learn a little bit too. But something was missing. I didn't feel the same level of intellectual excitement that I had felt in college. There was no Ultimate Paradox, no obsession driving my work. I began to recognize the importance and value of having obsessions.

Then, in 1982, I wrote a long article (a cover story for *Business Week* magazine) about research in the field of artificial intelligence. I talked with many leading researchers in the field. I became increasingly interested in questions about the mind. How can a mind emerge from a collection of mindless parts? It seems clear that no one part is "in charge" of the mind (or else it too would be a mind). But how can a mind function so effectively and creatively without anyone (or anything) in charge?

At last, I had a new Ultimate Paradox, a new obsession. I wasn't so much interested in the details of neuroscience, or even in the traditional research of artificial intelligence. Rather, I wanted to develop qualitative ways to think about my new Ultimate Paradox. I became interested not only in minds but also in other systems in which simple parts organize themselves into complex and sophisticated wholes. I became interested in evolution, hoping to gain a better understanding of how today's sophisticated life forms evolved from a few simple chemicals. For me, there was something deeply intriguing, and deeply beautiful, about this self-organized emergence of order from disorder, of complexity from simplicity. I developed a strong emotional investment in this idea. Few things got me more upset than listening to creationists attacking the idea of evolution, attacking the idea that complexity can arise, on its own, from simple pieces.

My new Ultimate Paradox led me to new questions and new challenges. I wanted to understand how people think about the organized patterns and structures they see in the world, and why they resist certain ways of thinking about them. How do people come to terms with the Ultimate Paradox of systems that organize themselves? How could I help people develop new ways of thinking about—and appreciating—such systems?

This book is a result of my decade-long obsession with my new Ultimate Paradox.

Acknowledgments

In 1983 I read three books that changed my life: *Mindstorms* (by Seymour Papert), *Structure and Interpretation of Computer Programs* (by Hal Abelson and Gerry Sussman), and *Gödel, Escher, Bach* (by Douglas Hofstadter).

Taken together, these books profoundly changed the way I think about children, minds, learning, and computers. A decade later, these three books remain the most influential (and exciting) books I have ever read. Virtually every page of this book was influenced, in one way or another, by ideas from those three books.

I was lucky enough to have two of the authors of those books as my advisers and mentors during my years of graduate study at MIT. The two of them—Hal Abelson and Seymour Papert—formed a wonderfully complementary pair. Together, they provided me with a full range of inspiration, encouragement, advice, and support. I came to MIT with a somewhat untraditional background, having worked as a journalist for the previous six years. But, from the very beginning, Seymour and Hal treated me as a serious colleague and made me feel at home at MIT. Over the years, each of them, in his own way, has deeply affected the way I think— and what I think about. I will be forever grateful to them.

Many other people contributed in many ways as my ideas grew into a doctoral dissertation and then into this book. In particular, I'd like to thank

- all of the high-school students who volunteered to participate in this project;

- Brian Silverman, for always being willing (no, eager) to brainstorm about Logo, mathematical and scientific ideas, and just about anything else;

- Randy Sargent, for implementing versions of StarLogo on UNIX machines and the Macintosh, and for suggesting many ideas about the design of StarLogo;

- Danny Hillis, for creating the computer that made it possible for me to create the first version of StarLogo;

• Alan Ruttenberg and David Sheppard, for helping me get started on the Connection Machine (and helping me when things went wrong);

• JP Massar and Mario Bourgoin of Thinking Machines Corp., and Ryan Evans and Andy Begel of MIT, for helping with the implementation of StarLogo on the Connection Machine;

• Uri Wilensky, for exploring new uses of StarLogo, and for sharing ideas about the role of StarLogo in education and learning;

• Steve Ocko, Fred Martin, and Randy Sargent, for collaborating on LEGO/Logo projects, and for making the LEGO/Logo Lab a fun place to be;

• Mike Eisenberg and Franklyn Turbak, for sharing countless ideas and jokes, and for making Tech Square a fun place to be;

• Robbie Berg, Ron Kahn, Cliff Rechtschaffen, and Natalie Rusk, for being very special friends;

• Chris Langton, for organizing (and inviting me to) the Artificial Life conferences;

• Edith Ackermann, for helping me to think about children and learning in new ways;

• Sherry Turkle, for giving me interesting things to think about whenever we talk;

• Andy diSessa, for serving as an informal adviser and mentor during (and after) my years in graduate school;

• David Chen and Kevin McGee, for making comments on a draft of my doctoral dissertation;

• members of the Epistemology and Learning (E&L) Group at the MIT Media Lab, for creating a stimulating (if sometimes fractious) environment in which I could develop and share new ideas;

• members of the Switzerland Group at Tech Square, for always being willing to share their ideas and technical expertise;

• Robert Rasmussen, Lars Bo Jensen, and Allan Toft, for making my interactions with the LEGO Group so enjoyable and productive;

• Wanda Gleason, Mai Cleary, and Jacqueline Karaaslanian, for providing administrative support;

• Nicholas Negroponte, director of the Media Lab, for creating a research environment where interesting ideas can grow and flourish;

• Betty and Harry Stanton, Jenya Weinreb, and others at MIT Press, for encouraging and helping with the publication of this book;

• my mother and father, for always being supportive and proud of everything I've ever done; and

• the LEGO Group, the National Science Foundation (Grants 851031-0195, MDR-8751190, and TPE-8850449), and the General Electric Foundation for providing financial support.

Turtles, Termites, and Traffic Jams

Organization of This Book

This book is divided into five chapters: Foundations, Constructions, Explorations, Reflections, and Projections.

Foundations provides background and framework for the rest of the book. It explores examples of decentralization in many different domains, as evidence of the breadth and depth of the trend toward decentralization.

Constructions describes the design of the StarLogo programming language. It discusses both the educational ideas (in particular, constructionism) and the computational ideas (in particular, massive parallelism) underlying the design of the language.

Explorations describes nine StarLogo microworlds—simulated worlds that highlight essential ideas about decentralization and self-organization. It discusses how high-school students and I created and explored these microworlds, and what we learned as we did.

Reflections analyzes ways of thinking about decentralization and self-organization, based largely on experiences with StarLogo microworlds. It probes the nature of the centralized mindset, discusses the allure of decentralization, and proposes heuristics for thinking about decentralized systems.

Projections looks ahead.

1

Foundations

Any study which throws light upon the nature of "order" or "pattern" in the universe is surely nontrivial.

—Gregory Bateson, *Steps to an Ecology of Mind*

Introduction

A flock of birds sweeps across the sky. Like a well-choreographed dance troupe, the birds veer to the left in unison. Then, suddenly, they all dart to the right and swoop down toward the ground. Each movement seems perfectly coordinated. The flock as a whole is as graceful—maybe more graceful—than any of the birds within it.

How do birds keep their movements so orderly, so synchronized? Most people assume that birds play a game of follow-the-leader: the bird at the front of the flock leads, and the others follow. But that's not so. In fact, most bird flocks don't have leaders at all. There is no special "leader bird." Rather, the flock is an example of what some people call "self-organization." Each bird in the flock follows a set of simple rules, reacting to the movements of the birds nearby it. Orderly flock patterns arise from these simple, local interactions. None of the birds has a sense of the overall flock pattern. The bird in front is not a leader in any meaningful sense—it just happens to end up there. The flock is organized without an organizer, coordinated without a coordinator.

Bird flocks are not the only things that work that way. Ant colonies, highway traffic, market economies, immune systems—in all of these systems, patterns are determined not by some centralized authority but by local interactions among decentralized components. As ants forage for food, their trail patterns are determined not by the dictates of the queen ant but by local interactions among thousands of worker ants. Patterns of traffic arise from local interactions among individual cars. Macroeconomic patterns arise from local interactions among millions of buyers and sellers. In immune systems, armies of antibodies seek out bacteria in a systematic, coordinated attack—without any "generals" organizing the overall battle plan.

In recent years, there has been a growing fascination with these types of systems. Ideas about decentralization and self-organization are spreading through the culture like a virus, infecting almost all domains of life. Increasingly, people are choosing decentralized models for the organizations and technologies that they construct in the world—and for the theories that they construct about the world.

Almost everywhere you look these days, there is evidence of decentralization. You can see it every time you pick up a newspaper. On the front page, you might see an article about the failure of centrally planned economies in Eastern Europe. Turn to the business page, and you might find an article about the shift in corporate organizations away from top-down hierarchies toward decentralized management structures. The science section might carry an article about decentralized models of the mind, or maybe an article about distributed approaches to computing. And in the book review you might read an article suggesting that literary meaning itself is decentralized, always constucted by readers, not imposed by a centralized author.

But even as the influence of decentralized ideas grows, there is a deep-seated resistance to such ideas. At some deep level, people seem to have strong attachments to centralized ways of thinking. When people see patterns in the world (like a flock of birds), they often assume that there is some type of centralized control (a leader of the flock). According to this way of thinking, a pattern can exist only if someone (or something) creates and orchestrates the pattern. Everything must have a single cause, an ultimate controlling factor. The continuing resistance to evolutionary theories is an example: many people still insist that someone or something must have explicitly designed the complex, orderly structures that we call Life.

This assumption of centralized control, a phenomenon I call the *centralized mindset,* is not just a misconception of the scientifically naive. It seems to affect the thinking of nearly everyone. Until recently, even scientists assumed that bird flocks must have leaders. It is only in recent years that scientists have revised their theories, asserting that bird flocks are leaderless and self-organized. A similar bias toward centralized theories can be seen throughout the history of science.

Of course, centralized ideas are not always bad or wrong. Some phenomena are described quite well by centralized theories. In some systems, there *are* leaders. And when people try to construct new technologies and new organizations, centralized strategies are often very useful. Sometimes it is a good idea to put someone or something in charge. The problem is that people have, too often, relied almost entirely on centralized strategies. Decentralized approaches have been ignored, undervalued, and overlooked. Centralized solutions have been seen as *the* solution.

That is starting to change, but only slowly. There is a powerful tension. On one side is the growing fascination with decentralized systems and self-organizing behaviors. On the other side is the deep commitment to centralized ways of thinking.

In this book I explore both the allure of decentralization and the centralized mindset that resists it. I examine how people think about decentralized systems and how they might learn to think about them in new ways. I describe new tools and activities that I designed to encourage people to experiment with new types of systems—and to engage in (and reflect upon) new types of thinking.

My investigation consists of several interwoven threads, each of which reinforces and enriches the others:

• *Probing people's thinking.* How do people think about self-organizing behaviors? To what extent do they assume centralized causes and centralized control, even when none exists? Are people even aware of such assumptions? In the cognitive science community, there has been a great deal of research into "folk physics," examining how people think about concepts from Newtonian physics. Here, I am interested in "folk systems science," aiming to understand how people think about systems.

• *Developing new conceptual tools.* In recent years, there has been considerable research into analytic techniques for describing and "solving" decentralized problems, and making accurate predictions about decentralized systems. But that is not my primary interest. Rather, I am interested in developing heuristics and qualitative tools to help people think about decentralized systems in new ways. My hope is that these conceptual tools will help people move beyond the centralized mindset.

• *Developing new computational tools.* Probably the best way to develop better intuitions about decentralized systems is to construct and "play with" such systems. To make that possible, I developed a massively parallel programming language that lets people control the actions of (and interactions among) thousands of computational objects. The language, called StarLogo, is an extension of Logo, a programming language commonly used in precollege education. Whereas traditional versions of Logo allow users to control a single graphic "turtle" (or maybe a few graphic turtles), StarLogo gives users control over thousands of graphic turtles. With StarLogo, people can create and explore a wide variety of decentralized systems. For example, a user might write simple programs for thousands of "artificial ants," then watch the colony-level behaviors that arise from all of the interactions.

High-school students have used StarLogo to create and explore a variety of decentralized microworlds. One pair of students programmed the motion of cars on a highway, exploring how and why traffic jams form. Another student used StarLogo to construct and explore an ecological system with turtles and grass. My observations of the students, along with self-observations of my own StarLogo projects, provided me with ideas for improving StarLogo as a language—and, more important, insights into how people think (and how, given new tools, they *might* think) about decentralized systems.

This research might seem like a strange mixture. What field is it in? Is it education? Computer science? Psychology? Epistemology? Biology? In my view, it is all of these—and necessarily so. It would be counterproductive to separate one from the others. Only by drawing on all of these domains is it possible to do justice to any of them.

The Era of Decentralization

On December 7, 1991, Russian president Boris Yeltsin met with the leaders of Ukraine and Belarus in a forest dacha outside the city of Brest. After two days of secret meetings, the leaders issued a declaration: "The Union of Soviet Socialist Republics, as a subject of international law and a geopolitical reality, is ceasing its existence." With that announcement, Yeltsin and his colleagues sounded the final death knell for a centralized power structure that had ruled for nearly 75 years. In its place, the leaders established a coalition of independent republics, and they promised a radical decentralization of economic and political institutions.

The next day, halfway around the world, another powerful institution announced its own decentralization plans. IBM chairman John Akers publicly announced a sweeping reorganization of the computer giant, dividing the company into more than a dozen semiautonomous business units, each with its own financial authority and its own board of directors. The goal was to make IBM more flexible and responsive to the needs of rapidly changing markets. As *Business Week* magazine put it, "The reorganization could amount to no less than a revolution in the way IBM does business."

Thus, within days, two of the world's most powerful institutions announced radical transformations, abandoning centralized hierarchies in favor of more decentralized structures. Of course, the reorganizations of the Soviet Union and IBM were not directly related to one another. But the two reorganizations are both part of a broad trend that is sweeping through our culture. Throughout the world, there is an unprecedented shift toward decentralization.

The decentralization trend is evident in the ways that people organize countries and corporations, and in the ways people design new technologies. But more important, it is evident in the ways people *think about* the world. More so than ever before, scientists are using decentralized models and metaphors to describe the phenomena they observe in the world. Increasingly, scientists (and others) are seeing decentralization wherever they look. It seems fair to say that we have entered an Era of Decentralization.

Of course, interest in decentralization is not entirely new. More than two hundred years ago, Adam Smith made a forceful argument against centralized government control of the economy. In *The Wealth of Nations,* published in 1776, Smith advocated decentralized markets as a more orderly and more efficient alternative to centralized control. He used the image of an "invisible hand" to drive home the radical idea that economic order and justice can be achieved (and, in fact, are more likely to be achieved) without centralized control of the economy. Each individual in a society, wrote Smith, "neither intends to promote the public interest, nor knows how much he is promoting it . . . he intends only his own gain, and he is in this, as in many other cases, led by an invisible hand to promote an end which was no part of his invention." This faith in the decentralized actions of individuals can also be seen in other political and philosophical writings of Smith's era—including the United States Declaration of Independence, written just a few months after the publication of *The Wealth of Nations.*

Nearly a century after Adam Smith, Charles Darwin brought the idea of the invisible hand to biology. Darwin's challenge was to explain the organized complexity of living systems. Even the simplest creatures of the living world are more complex than the most complex machines of the technological world. Who or what is responsible for this organized complexity of living systems? Before Darwin, nearly everyone accepted a centralized explanation: God designed the complexity of creatures. In *Origin of Species,* Darwin offered the first serious alternative: his (decentralized) theory of natural selection. Just as Adam Smith asserted that centralized government control is not needed to create order in the economy, Darwin asserted that a centralized designer of life is not needed to create order in the living world. Instead, order and complexity arise from the decentralized processes of variation and selection.

So interest in decentralization is not a new phenomenon. But there *is* something new and different today. Ideas about decentralization are now spreading more widely, and penetrating more deeply, than ever before. More people are open to the idea of decentralization. Decentralized phenomena have a high salience in today's culture: they are attracting more attention, generating more interest. As a result decentralization has emerged as a theme in almost every domain of human activity. We seem to be undergoing a revolutionary change—what Thomas Kuhn would call a "paradigm shift"—in the way we see and construct the world.

This section examines the trend toward decentralization in five different domains:

• Decentralization in organizations

• Decentralization in technologies

• Decentralization in scientific models

• Decentralization in theories of self and mind

• Decentralization in theories of knowledge

As I investigated the growing interest in decentralized ideas in so many varied domains, my first inclination was to try to figure out which domain is the most influential. Does one of these domains act as the primary catalyst of decentralization, sparking decentralization in other domains? Perhaps new decentralized scientific models are influencing the ways we design our organizations and technologies? Or maybe it is the decentralization of technology that is provoking us to view the natural world in more decentralized ways?

But as I thought about it, I realized that my inquiry was violating the spirit of the very trend that I was trying to study. Why should there be a single, central, underlying cause for all of this decentralization? It seems better to view these domains as a type of auto-catalytic system: the decentralization of each domain reinforces and catalyzes the decentralization of the others. Most likely, there is no single, ultimate cause. Each domain provides new models and new metaphors that influence the others, refining and accelerating the decentralization trend.

The following overview is necessarily superficial, ignoring many of the subtleties and exceptions to the decentralization trend. It paints in broad strokes, not fine detail. Its goal is to provide the big picture of how decentralized ideas are spreading through the culture, affecting nearly all domains of life.

Decentralization in Organizations

The spread of decentralized ideas can be seen in organizations of all sizes and types—countries, companies, schools, clubs. Although details are different in each case, the basic idea is always the same: pushing authority and power down from the top, distributing rights and responsibilities more widely.

For some countries (such as the Soviet Union) decentralization has meant breaking apart into separate pieces. But changes in national boundaries are not nearly as important as changes in political and economic structures. Politically, countries throughout the world are shifting away from totalitarianism toward democracy. Economically, countries are shifting away from centrally controlled economies toward market-oriented economies. As a result, decision making (both political and economic) is becoming more decentralized than ever before.

Of course, there are exceptions to the trend. In China, the government reasserted its centralized power with the brutal crackdown in Tiananmen Square. And in many of the former Soviet republics, democracy is very fragile. But the overall trend is clear. Between 1989 and 1991, countries with a combined population of 1.5 billion people, more than one-quarter of the world's population, moved away from autocratic toward more democratic forms of government, according to Freedom House, an American human-rights group. Now, for the first time ever, more than half of all countries are democracies.

A growing faith in market mechanisms is an important component of the decentralization trend. Many countries that previously relied on centrally planned economies are now switching to market-oriented approaches. And countries where market-based economics are already firmly entrenched are starting to use market mechanisms even more than before. In the United States, the government is increasingly using market mechanisms as part of the regulatory process. In the past, the Federal Communications Commission decided how to allocate frequencies on the radio spectrum. But the commission recently proposed a new approach: let new spectrum users (for example, wireless telephones) buy frequencies from existing users (for example, microwave communications by railroads). Similarly, the government is now allowing companies to buy and sell "rights to pollute." Each factory has pollution guidelines. But it can exceed those guidelines if it buys "pollution credits" from another factory that keeps its own pollution levels sufficiently below the guidelines.

In American education, decentralization is playing a role on several levels. The school-choice movement brings market-oriented thinking to the world of education, asserting that individual families—not the government—should decide where children go to school. Meanwhile, another movement called school-based management is pushing for a different type of decentralization: shifting decision-making authority from district (and state) offices to individual schools. Inside the classroom, a growing number of educators are recognizing the value of child-centered approaches to learning, transforming the teacher from a central authority figure into a catalyst, coach, and collaborator.

In the corporate world, too, there is decentralization on several levels. The rise of entrepreneurship in the 1980s led to a proliferation of small companies and independent consultants. That trend is likely to continue. Economic activity can be coordinated in two different ways: either a company makes the parts it needs internally (via vertical integration), or it buys parts from outside suppliers (via the market). For example, General Motors can make its own tires, or buy them from Goodyear. In the past, the high "coordination costs" of external purchases led many companies to make parts internally. But improvements in information technology are decreasing coordination costs, shifting the balance toward greater use of outside markets—and, thus, a proliferation of smaller firms (Malone, Yates, and Benjamin 1987).

At the same time, management structures within companies are also becoming decentralized. Since the beginning of the Industrial Revolution (and even before), companies have organized themselves as pyramid-like hierarchies. Information flowed up the hierarchy to the top, where decisions were made and passed back down the hierarchy. Thus, power, authority, and decision making were centralized at the top in most corporations—and in many other organizations that followed the corporate model.

That is now changing. A 1989 *Harvard Business Review* article called "Managing without Managers" explains: "The organizational pyramid is the cause of much corporate evil, because the tip is too far from the base. Pyramids emphasize power, promote insecurity, distort communications, hobble interaction, and make it difficult for the people who plan and the people who execute to move in the same direction" (Semler 1989). In place of the traditional pyramid, companies are "flattening" their organizational structures by getting rid of middle managers and distributing decision-making responsibility more evenly through the organization. The movement started with employee participation in "quality circles" in the 1970s. Now companies are giving workers more responsibilities over production decisions. Some are even experimenting with "self-management teams"—that is, teams without bosses (Dumaine 1990). Someday, companies could end up with what MIT sociologist Charles Sabel calls a "Mobius strip organization"—an organization without a top or bottom.

Decentralization in Technologies

The decentralization in organizational structures is linked, in part, to decentralization of technologies. This connection was particularly apparent during the attempted Soviet coup in 1991, when hard-liners tried to reassert centralized control. As John Barlow (1992) wrote, "Because of the decentralized and redundant nature of digital media, it was impossible for the geriatric plotters in the Kremlin to suppress the delivery of truth. Faxes and email messages kept the opposition more current with developments than the KGB, with its hierarchical information systems, could possibly be."

Computer technologies have not always been viewed as a decentralizing force. Just 30 years ago, computers were synonymous with centralized power. Only the largest institutions could afford computers. And within those institutions only a few privileged people had direct access to the machines. To run a program, you had to deliver a stack of cards (or tape) to a member of the "computer priesthood" that guarded and cared for the machine. Not surprisingly, college students in the 1960s saw computers as impersonal tools used by the Establishment to keep control over the masses.

But as the cost and size of electronics continued to decline, the uses (and perceptions) of computers changed radically. In the 1970s time-sharing technology gave more people access to computers. To run a computer program, you could sit at a terminal (maybe on your own desk) and interact with the computer in real time. But the computer itself was still centralized and shared. The real breakthrough came with the personal computers of the 1980s. Suddenly computers began to appear on desks everywhere. In 1972 there were only 150,000 computers. A decade later there were several million computers. Today there are more than 100 million computers.

The decentralization trend continues today with the proliferation of notebook computers and even palmtop computers. Computers are becoming part of the environment itself, invisibly buried within all types of objects (such as televisions, fax machines, and telephones). Ultimately all of these objects will be linked together, in a decentralized computational web.

Even as computers spread through offices, factories, and homes, most computers remain quite centralized in their internal architecture. Most of today's computers continue to use an architecture developed by John von Neumann nearly half a century ago. This von Neumann architecture is based on a single "central processing unit" that performs and organizes most of the computational work. All information must flow through that single processor.

But that too is changing. A growing number of companies are developing parallel computers—computers with more than one processor inside. Some "massively parallel" computers have tens of thousands of processors, and there are plans for computers with more than a million processors. With a parallel computer, a user can divide a problem into many separate parts, then assign different processors to work on different parts of the problem at the same time. The challenge is to find ways for all of the processors to remain coordinated—just as birds remain coordinated within a flock.

Thus the decentralization of computation proceeds at multiple levels, in an almost fractal-like fashion. As computational power becomes decentralized throughout society, it is also becoming decentralized within the computers themselves.

Decentralization in Scientific Models

For three hundred years, the models and metaphors of Newtonian physics have dominated the world of science—and, even more so, people's perceptions of science. Newton offered an image of the universe as a machine, a clockwork mechanism. Newton's universe is ruled by linear cause and effect—one gear turns, which makes a second gear turn, which makes a third gear turn, and so on. This cause-effect relationship is captured in Newton's famous $F=ma$ formula: force gives rise to acceleration; cause gives rise to effect.

In the common perception of the Newtonian universe, the idea of "mutual interaction" is de-emphasized. When people think of interactions in the Newtonian universe, they think of one object acting on another. One object acts as the cause, the other object receives or suffers the effect. One object is in control, the other is acted upon. Most of the attention goes to Newton's first two laws of motion, which focus on how a force influences the motion of an object. Much less attention goes to Newton's third law, which focuses on the reaction that accompanies every action.

During the twentieth century, the Newtonian view of the world has been challenged on many different fronts. One of the most serious challenges comes from the growing interest in so-called complex systems. In an increasing number of fields, scientists have shifted metaphors, viewing things less as clocklike mechanisms and more as complex ecosystems. Rather than viewing the world in terms of one individual object acting on another in a neat causal chain, researchers are viewing the world in terms of decentralized interactions and feedback loops. They are studying how complex behaviors can emerge from interactions among simple rules, and how complex patterns can emerge from interactions among simple components.

This growing interest in "emergent" phenomena has been accompanied by confusion and controversy, since different people use the term *emergent* in different ways. Many popular descriptions of emergence (and even some scientific ones) are tinged with mysticism, as if something magical is going on. But no magic is needed. As I am using the term, emergence is fully consistent with most traditional scientific ideas— including Newtonian physics. The point is not that Newtonian models are *wrong*. It is that Newtonian models are *inappropriate* for trying to make sense of certain types of phenomena. New types of models are needed, operating at a different "level" from Newtonian models, focusing on the behaviors of systems, not the actions of individuals.

Many ideas about emergence and complexity have been inspired by research in the biological fields of ecology, ethology, and evolution. In one classic study, ethologist Niko Tinbergen described how the behavior of a stickleback fish emerges from interactions among several simple rules. And much ecological research looks at how large-scale patterns emerge from local interactions among living organisms. But interest in complex systems has spread far beyond biology. In the 1940s and 1950s the field of cybernetics attempted to create a new unifying framework for understanding all types of systems in the world—be they biological, social, or technological. The field attracted engineers, biologists, psychologists, anthropologists, and others. Working together, researchers tried to find and forge connections among their disciplines, looking for similarities in the behaviors of minds, machines, animals, and societies.

While cybernetics never developed into a mainstream discipline, many of its core ideas (such as feedback and self-organization) are alive and increasingly influential in scientific thinking. Nobel Laureate Ilya Prigogine and his associates have shown how physical and chemical systems can exhibit the same types of self-organizing behaviors that are typical of biological systems. Under the right conditions, for example, a heated liquid will form rotating "convection cells," where the rotating patterns are millions of times larger than the range of the intermolecular forces that cause them. Similarly, certain chemical reactions (such as the Belousov-Zhabotinski reaction) can exhibit either spatial patterns (large-scale spirals) or temporal patterns (periodically changing color). Self-organizing patterns can also arise in technological systems. Computers interconnected on networks can sometimes behave as "computational ecologies": interactions among individual machines can give rise to surprising network-wide patterns (Huberman 1988).

The study of self-organizing systems is one strand in the study of nonlinear dynamical systems, the rapidly growing research effort that aims to find common mathematical foundations for all types of complex behavior. Dynamical systems that exhibit *chaotic* behavior have received a particularly high level of attention and publicity in recent years. As noted by Farmer and Packard (1986), the study of self-organizing systems is, in some ways, the "related opposite" of the study of chaos: in self-organizing systems, orderly patterns emerge out of lower-level randomness; in chaotic systems, unpredictable behavior emerges out of lower-level deterministic rules.

The new field of artificial life is a striking example of the growing interest in self-organization and decentralized scientific models. Artificial life researchers aim to gain a better understanding of living systems by creating computational versions of them—for example, creating artificial versions of ant colonies or bird flocks. In their efforts, artificial life researchers are guided by an abiding faith in decentralized approaches. As Chris Langton (1989) wrote in the founding article of the new field,

The most promising approaches to modeling complex systems like life or intelligence are those which have dispensed with the notion of a centralized global controller, and have focused instead on mechanisms for the *distributed* control of behavior. . . . Artificial Life studies natural life by attempting to capture the behavioral essence of the constituent components of a living system, and endowing a collection of artificial components with similar behavioral repertories. If organized correctly, the aggregate of artificial parts should exhibit the same dynamic behavior as the natural system. This bottom-up modeling technique can be applied at any level of the hierarchy of living systems in the natural world—from modeling molecular dynamics on millisecond time-scales to modeling evolution in populations over millenia.

Decentralization in Theories of Self and Mind

Few things seem more obvious than the singular nature of the mind and self. Each of us experiences life as a single thread of consciousness. Each of us feels as if we have a single, unified presence in the world. In the words of Francisco Varela and his associates, each of us has "a stable and constant vantage point from which to think, perceive, and act" (Varela, Thompson, and Rosch 1991). Each of us imagines our own mind as "I," not "we."

But the idea of the unified, centralized mind has eroded during the past century—and the erosion has accelerated in the past decade. The beginnings of the "decentering" of self and mind can be seen in the nineteenth-century writings of Sigmund Freud. Freud's "unconscious" was a direct attack on the idea of a single executive in charge of the mind. Freud saw the unconscious as an equal participant (with the conscious) in the workings of mind. The unconscious, according to Freud, is not a passive repository of forgotten ideas but a lively agent actively repressing thoughts. Freud further fragmented the mind with his formulation of the ego, the superego, and the id—with the superego and id pulling the ego in different directions.

The decentralized nature of Freud's theories met with resistance, even among Freud's followers in psychoanalytic research. So-called ego psychologists focused their attention on the ego, viewing the ego as a type of leader or chief executive within the mind. Ego psychology, writes Sherry Turkle (1988), "takes what is most subversive [in Freud's theories]—the decentered self—and softens it."

In recent years, however, psychoanalytic research has swung back toward decentralized models, particularly with the rise of object relations theory. This theory, exemplified by the work of Melanie Klein, describes psychological development in terms of the "internalization" of objects. Relationships with people in the world are internalized as agents or objects within the mind. Freud took a step in this direction, describing the superego as the internalization of the ideal parent. But object relations theory goes much further, proposing an entire society of inner agents within the mind. The self emerges from the interactions among the internalized objects.

The field of artificial intelligence (AI) studies the mind from a perspective very different from that of psychoanalysis. But AI too has moved toward more decentralized models in the last decade. The early days of AI, in the 1950s, were characterized by a diversity of approaches. Some researchers experimented with perceptrons and neural networks, networks of simple computational elements. No single element was in charge of the network. Rather, solutions emerged based on interactions among the distributed elements. By contrast, Newell and Simon's classic General Problem Solver (GPS) represented a much more centralized approach. GPS acted more like a single computational agent, solving problems by continually reducing the distance between its current state and its goal state.

By the mid-1960s centralized approaches had become dominant in AI. A great deal of research focused on planning systems such as Strips (Fikes and Nilsson 1971). Such systems typically tried to construct a single, coherent model of the world, then planned a course of action based on that model. In expert systems, another popular line of research, a centralized "inference engine" made deductions and decisions based on a "knowledge base" of rules, often culled from human experts.

For 20 years such centralized approaches reigned supreme. Then, in the mid-1980s, research in AI began to shift. There was a renewed enthusiasm for neural networks. The book *Parallel Distributed Processing* (Rumelhart, McClelland, and the PDP Research Group 1986) became a bible for a new generation of neural-network researchers. Researchers argued, almost mystically at times, that high-level symbolic representations would emerge from interactions among "subsymbolic" elements in their networks.

AI research is now filled with decentralized models. In the "Society of Mind" model (Minsky 1987) societies of mental agents work together (and compete with one another) to do things that no agent could do on its own. In the "subsumption architecture" (Brooks 1991) robots are controlled by collections of simple "behavior modules" (such as *wander* and *avoid* and *explore*), each of which responds to sensory inputs from the world. A robot's behavior emerges from interactions among these modules (and their interactions with the world). Recently, philosopher and cognitive scientist Daniel Dennett (1991) has proposed a "multiple drafts" model of consciousness, arguing that there is no single stream of consciousness in the mind. Rather, multiple narratives are simultaneously created and edited in different parts of the mind/brain. The idea of a single stream of consciousness, he argues, implies a "single functional summit or central point" where it all comes together. And such a summit does not exist. "The idea of a special center in the brain is the most tenacious bad idea bedeviling our attempts to think about consciousness," writes Dennett.

Current models of the mind differ significantly from one another. For example, neural networks are based on an architecture of simple, homogeneous, highly connected components, while the Society of Mind is based on an architecture of complex, specialized, semi-insulated subsystems. Nevertheless, these disparate models are all far more decentralized than the models that dominated AI research in the 1970s and early 1980s.

Are the new decentralized views of the mind correct? That is not the issue here. What is important (and undeniable) is that decentralized models have captured the imaginations of researchers studying the mind. Why the shift? Without a doubt, the other technological and intellectual trends described in this chapter have been an important influence, creating an environment in which decentralized models of mind seem natural and sensible. And conversely, these new images of the mind are undoubtedly influencing the ways people think about everything else.

Decentralization in Theories of Knowledge

Our theories of the world and of ourselves are becoming increasingly decentralized. But that is not all. Decentralization is also happening at a meta-level: our theories about theories—or, more generally, our theories about knowledge—are also becoming decentralized.

For centuries, philosophers strived for "objective knowledge." They put great faith in the power of logic to systematize all knowledge, to find ultimate "meaning" and "truth." But belief in the existence of objective knowledge began to unravel in the 1930s and 1940s, spurred (in part) by new scientific and mathematical theories, such as Heisenberg's uncertainty principle and Gödel's theorem. Today philosophers continue to move away from the notion of a single, absolute, unifying conception of knowledge, arguing instead that knowledge is constantly constructed and reconstructed in a much more decentralized way.

Current trends in literary criticism serve as an example. Traditional theories of literature assumed that meaning was created by an author and conveyed through the author's writings. According to this view, reading is a search for inherent meaning in a document, an attempt to decipher the intention of the author. But modern schools of literary criticism—such as poststructuralism, reader-response theory, and deconstructionism—adopt a very different stance. These movements all focus on readers (not authors) as the main constructors of meaning. In this new view, texts have little or no inherent meaning. Rather, meanings are constantly reconstructed by communities of readers through their interactions with the text. Meaning itself has become decentralized.

Literary scholar Alvin Kernan discusses (and criticizes) this shift in literary criticism in his book *The Death of Literature* (1990). Kernan describes the new literary trends in a somewhat cynical tone—but in terms that proponents of the new trends would probably embrace: "It would not be an exaggeration to say that the *nouvelle critique* has been an egalitarian revolt against the authority of the writer and the book in the name of the reader and reading. Where once authors were considered geniuses inscribing truth in great works of art, they now are declared to be dead. Where great books were thought to embody precise and autonomous meanings, radical critics have declared books empty of meaning in their own right. Freed of the restraints of 'authoritarian' authors and texts that control their own meanings, so the argument goes, readers may now give their own interests and imaginations free rein."

Another challenge to traditional theories of knowledge comes from feminist scholarship. One of the central ideas of feminist studies is the existence of many different, but equally valid, ways of knowing and thinking—multiple "voices," in the words of Carol Gilligan (1982). Since different voices are often valued differently in our society, the idea of multiple voices takes on a political edge. Formal, logical, abstract, and analytical ways of thinking, typically associated with men, have been privileged in our society—viewed as superior, more advanced, more likely to lead to "truth." On the other hand, relational and contextual ways of thinking, typically associated with women, have been undervalued and discouraged.

In a now-classic example Carol Gilligan describes a hypothetical moral dilemma that she posed to two 11-year-old children, Amy and Jake. In the dilemma, a man is trying to obtain drugs for his dying wife, but he doesn't have enough money to buy the drugs. Should the man steal the drug? Jake sees the dilemma as "sort of like a math problem with humans." He applies a mathematics-type logic: human life is worth more than money, so the man should steal the drug. Amy, on the other hand, resists a neat solution to the dilemma. She worries about how the theft of the drug might affect the relationship between the man and his wife. She suggests that the man talk to the druggist, believing that if they "talked it out long enough, they could reach something besides stealing."

According to traditional research on moral development, Jake's response is seen as more "advanced" than Amy's. Using Lawrence Kohlberg's (1981) six-stage theory of moral development, Amy ranks a full stage lower than Jake. But Gilligan argues that Amy's response is not inferior, only different. Whereas Jake focuses on a hierarchical system of rules and logic, Amy focuses on relationships and negotiation. The two approaches are complementary; developmentally, neither precedes the other. In her own way Amy is just as advanced as Jake.

Sherry Turkle and Seymour Papert (1990) point to feminist research like Gilligan's as one of several intellectual challenges to the "dominant epistemology" that grants privileged status to abstract, formal, and logical ways of thinking. They note that ethnographers of science are also challenging that dominant epistemology, finding that "scientific discoveries are made in a concrete, ad hoc fashion, and only later recast into canonically acceptable formalisms." In their own research on uses of computers, Turkle and Papert find "diversity in the practice of computing that is denied by its social construction" as a logical, formal activity. Some people, they note, form relationships with computation "more reminiscent of a painter than a logician." Equal access to computation, argue Turkle and Papert, requires an "epistemological pluralism"—an acceptance of the validity of multiple ways of knowing and thinking.

These challenges to the dominant epistemology resonate with the growing interest in decentralization in several ways. For one thing, alternative epistemologies typically have a decentralized feel to them: they are based on ideas such as relationships and interdependencies, not logical hierarchies. But probably the most compelling aspect of alternative epistemologies is not in the *nature* of those epistemologies but in the very *idea* of alternative epistemologies—the idea that there are multiple ways of thinking and knowing. People are recognizing that knowledge speaks not with a single voice but with many.

Looking Ahead: From Foundations to Constructions

The growing interest in decentralization means much more than new types of organizations. It means new ways of viewing the world, new ways of thinking, and new ways of knowing. The rest of this book traces how people confront, resist, and make sense of these new ways of thinking and knowing. To what extent do people remain committed to centralized ways of thinking? How can new ways of thinking be nurtured and encouraged?

Next, in Constructions, I discuss new tools that people can use to construct, play with, and think about decentralized systems.

2

Constructions

To understand is to invent.

—Jean Piaget

Constructionism

As we enter the Era of Decentralization, there is an important educational challenge: How can we help people become intellectually engaged with the new types of systems and new types of thinking that characterize this new era? To date, schools and other educational institutions have done little, if anything, to engage students with the ideas of decentralization. Instead, they often perpetuate centralized explanations and approaches.

In a way, people have lots of experience with decentralized systems. They observe many decentralized systems in the world, and they participate in decentralized systems all the time. But observation and participation do not necessarily lead to strong intuitions or rich understanding. People observed flocks of birds for thousands of years before anyone figured out that flocks are, in fact, leaderless. And people participate in traffic jams without much understanding of the decentralized interactions that cause the jams. Observation and participation are not enough. People need a richer sense of engagement with decentralized systems. The best way to do that, I believe, is to give people opportunities to *design* decentralized systems.

This idea of learning through design is one aspect of what Seymour Papert has called the *constructionist* approach to learning and education (Papert 1991a). Constructionism involves two types of construction. First, it asserts that learning is an active process, in which people actively construct knowledge from their experiences in the world. (This idea is based on the *constructivist* theories of Jean Piaget.) To this, constructionism adds the idea that people construct new knowledge with particular effectiveness when they are engaged in constructing products that are personally meaningful. They might be constructing sand castles, LEGO machines, or computer programs. What's important is that they are actively engaged in creating something that is meaningful to themselves or to others around them.

Papert contrasts constructionism with instructionism. Whereas instructionism focuses on new ways for teachers to instruct, constructionism focuses on new ways for learners to construct. Both are important. But significant improvements in education are much more likely to come from advances in constructionism, not instructionism.

The major challenge for educators and educational developers, then, is to create tools and environments that engage learners in construction, invention, and experimentation. This process involves (at least) two levels of design: educators need to design things that allow students to design things.

In the next section, I focus on a particular constructionist tool: a programmable robotics system called LEGO/Logo. My goal in discussing LEGO/Logo is to provide a richer sense of the constructionist philosophy. In addition, I discuss the possibilities—and the limitations—of using LEGO/Logo to explore decentralized systems and self-organizing behaviors.

A Constructionist Contradiction?

The idea of a constructionist approach to the study of decentralized and self-organizing systems might seem, at first glance, like a contradiction. After all, how can you design a self-organizing phenomenon? By definition, self-organizing patterns are created without a centralized designer. But there are ways to use design in the study of self-organizing systems. Imagine that you could design the behaviors of lots of individual components—then observe the patterns that result from all of the interactions. For example, imagine you could design the behaviors of thousands of individual ants, then observe the colony-level patterns that result. This is a different sort of design: You control the actions of the parts, not of the whole. You are acting as a designer, but the resulting patterns are not designed. Thus the idea of a constructionist approach to learning about decentralization is not a contradiction.

LEGO/Logo

In 1985, I began working with Steve Ocko and Seymour Papert on a new type of construction set. We envisioned a construction set that would allow children to construct buildings and machines, as they had done for years with erector sets, Tinker Toys, and LEGO bricks. But in addition we wanted children to be able to program and control the things they built. After building a model house, a child should be able to add lights to the house, and program the lights to turn on and off at particular times. And the child should be able to add a garage, and program the garage door to open whenever a car approaches.

The idea was to combine several different design activities. Children would not only design architectural structures and gearing mechanisms, they would also design computer programs to control them. So we formed a collaboration with the LEGO toy company, and we began to link LEGO building bricks with the Logo programming language, a combination that we called LEGO/Logo.

Logo itself was developed in the late 1960s as a programming language for children (Papert 1980). In the early years, the most popular use of Logo involved a "floor turtle," a simple mechanical robot connected to

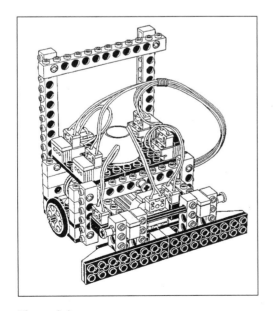

Figure 2.1
A LEGO Turtle

the computer by a long "umbilical cord." Logo included commands like `forward`, `back`, `left`, and `right` to control the floor turtle. For example, a child could type `forward 50` to make the turtle move forward by 50 "turtle steps," or `right 90` to make the turtle turn right through 90 degrees. The turtle makes possible a new approach to thinking about geometry, contrasting sharply with the Euclidean methods traditionally taught in the classroom. This new "turtle geometry" has proved to be much more intuitive for children. The turtle connects to children's experiences in the world—children can "play turtle," imagining themselves as the turtle. As a result the turtle has helped many children form a new relationship with mathematical ideas.

With the proliferation of personal computers in the late 1970s, the Logo community shifted its focus to "screen turtles." Children still use commands like `forward` and `right`, but these commands control graphic images of turtles on the computer screen, not actual mechanical robots. Screen turtles are much faster and more accurate than floor turtles, and thus allow children to create and investigate more complex geometric effects. Logo is now used in about one-third of the elementary schools in the United States.

In some ways, LEGO/Logo might seem like a throwback to the past, since it brings the turtle off the screen and back into the world. But LEGO/Logo differs from the early Logo floor turtles in several important ways. First of all, LEGO/Logo users are not given ready-made mechanical objects; they build their own machines before programming them. Second, children are not restricted to turtles. Elementary-school students have used LEGO/Logo to build and program a wide assortment of creative machines, including a programmable pop-up toaster, a "chocolate-carob factory" (inspired by Roald Dahl's children's stories about Willy Wonka), and a machine that sorts LEGO bricks according to their lengths (Resnick, Ocko, and Papert 1988). The LEGO toy company now sells a commercial version of LEGO/Logo. It is used in more than five thousand elementary and middle schools in the United States.

LEGO/Logo includes new types of LEGO blocks for building machines, and new types of "Logo blocks" for building programs. In addition to the familiar LEGO building bricks, there are new LEGO pieces like gears, pulleys, wheels, motors, lights, and sensors. There are optosensors that report when they detect changes in the level of light, and touch sensors that report when they are pressed.

As its programming language, LEGO/Logo uses an expanded version of Logo. The language includes new commands like `on` and `off` for controlling LEGO motors and lights, and new "reporter procedures" like sensor? for getting information from LEGO sensors. Just as students can build increasingly complex structures and machines by snapping together LEGO bricks, they can build increasingly complex computer programs by "snapping together" Logo commands. Imagine, for example, a LEGO car with a touch sensor on the front. A student can write a Logo program called go-until-bump that turns the car motor on, waits until the car bumps into something, then turns the car motor off. The program would look like this:

```
to go-until-bump
on
waituntil [sensor?]
off
end
```

A Sample LEGO/Logo Project

John, a fifth-grader at the Hennigan Elementary School in Boston, had an alarm clock next to his bed at home. But the alarm clock wasn't very effective. Often, when the alarm went off, John simply shut off the alarm and went back to sleep. John was determined to invent a better solution. His goal: to design an alarm clock that could not be ignored.

John started by playing with the LEGO optosensor. He placed the optosensor by the window, so the computer could "know" when the sun came up. But what should happen at sunrise? John had an idea. He built a small LEGO bed, with a small LEGO person on top. Underneath the bed he placed a hinged platform, so the bed could tilt from side to side. Alongside the bed he built a conveyor belt. Then he wrote a Logo program. When the optosensor detected light coming through the window, the program turned on two motors. One motor made the LEGO bed tilt to the side, making the LEGO person slide off onto the conveyor belt. The other motor turned the conveyor belt, carrying the LEGO person out the door.

Would John want a full-size version of his alarm–clock ejection bed for his home? Not really, he said. But he certainly enjoyed watching the little LEGO person fly out the door.

In recent years, the science education community has embraced the idea of "hands-on" education. LEGO/Logo is clearly hands-on: students have their hands on the LEGO bricks while building LEGO machines, and their hands on the computer keyboard while writing Logo programs. But hands-on is not enough. In many hands-on activities in school classrooms, students simply follow a list of instructions ("pour the liquid in test tube A into test tube B . . . "). Students are told what to do, and they do it. Their hands are on, but their heads are out.

The constructionist approach goes beyond hands-on in a variety of ways. In constructionist activities, students do not simply *manipulate physical objects,* they *construct personally meaningful products.* It is easy to see how "constructing" is better than merely "manipulating": children are sure to learn more by building and programming their own robots rather than manipulating store-bought, fully assembled robots. But there is a deeper point here. Children are likely to become intellectually engaged only if they are constructing *personally meaningful* things. When students design and construct products that are meaningful to themselves (or to others around them), they tend to approach their work with a sense of caring and interest that is missing in most school activities. In doing so, students are more likely to explore, and to make deep "connections" with, the mathematical and scientific concepts that underlie the activities. Building and programming a merry-go-round is based on the same underlying principles as building and programming a classic robot—but for a child who cares more about merry-go-rounds than robots, the merry-go-round project offers a much richer learning experience.

LEGO/Logo materials are designed to encourage such activities. LEGO bricks and Logo software are easy for children to relate to, and they are "plastic" enough (no pun intended) so that different students can use them in different ways, according to their own personal interests and obsessions. Of course, the materials alone are not enough. Like all materials, LEGO bricks and Logo software can be used in a rigid and formulaic way—and, in fact, they are used that way in (too) many classrooms. Such activities might be hands-on, but they are certainly not in the constructionist spirit. For constructionist activities to have any meaning, children must be given the freedom to follow their fantasies and the support to make those fantasies come to life. Building a chocolate-carob factory, inspired by Willy Wonka, is fundamentally different from building a conveyor belt, assigned by the teacher.

LEGO/Logo and Math

George, a third-grade student, began his LEGO/Logo work by building a simple LEGO car. First, he connected the car's motor to a battery box and watched the car roll forward. Next, he connected the motor to the computer and began experimenting with some of the new Logo commands. After a while, George put several commands together in the following expression:
repeat 4 [onfor 20 rd]

When George executed this expression, the computer turned on the motor for two seconds (onfor 20), reversed the direction of the motor (rd), then repeated those commands three more times. The result: the car moved forward and back, then again forward and back, completing two forward-back cycles.

Next, George changed the numerical input to repeat. He tried repeat 3 and repeat 6 and repeat 7. After this experimentation, George noticed a pattern: "When I use an even number, the car ends up where it began. When I use an odd number, it ends up away [from where it started]." George paused for a moment and then added, "So that's why there are even and odd numbers!"

Clearly, George had previously learned about even and odd numbers in the classroom. But George's experimentation with the LEGO car provided him with a new (and more personally relevant) representation of the concept. Moreover, the LEGO activity allowed George to relate to numbers in a new way: he played with the ideas of even and odd. This new relationship with even and odd numbers helped George develop a new level of understanding.

LEGO, Logo, and (Artificial) Life

Many LEGO/Logo projects involve classic centralized control: the computer tells one motor to turn on, then it tells the next motor to turn on, in a pre-planned sequence of actions. But LEGO/Logo can also be used to explore certain ideas about decentralized systems and self-organizing behaviors.

Consider, for example, a simple LEGO/Logo creature with a light sensor pointing upward (in the spirit of the "vehicles" described by Valentino Braitenberg (1984)). Say that the creature is programmed with two rules: (1) move forward when you detect light, (2) move backward when you are in the dark. When this creature is released in the environment, it goes forward until it moves into a shadow. Then it moves backward until it leaves the shadow. Then forward again. And so on, oscillating around the edge of the shadow. The creature can be viewed as an "edge-finding creature." This edge-finding capability is not explicitly represented in the creature's two rules. Rather, it is a type of "group behavior" that emerges from the interaction between the two rules—much as a flock's behavior emerges from interactions among the birds.

LEGO/Logo creatures provide a natural context for learning about *levels* in decentralized systems. Students tend to view their creatures on different levels at different times. Sometimes they view their creatures on a *mechanistic* level, examining how one LEGO piece makes another move. At other times they shift to an *information* level, exploring how information flows between the computer and LEGO motors and sensors. At still other times students view the creatures on a *psychological* level, attributing intentionality or personality to the creatures. One creature "wants" to get to the light. Another creature "likes" the dark. A third is "scared" of loud noises.

Often, students shift rapidly between levels of description. Consider the comments of Sara, a fifth-grade student at the Hennigan Elementary School in Boston. Sara was considering whether her LEGO/Logo creature would make a sound when its touch sensor was pushed: "It depends on whether the machine wants to tell . . . if *we* want the machine to tell us . . . if we *tell* the machine to tell us."

Within a span of ten seconds, Sara described the situation in three different ways. First she viewed the machine on a psychological level, focusing on what the machine "wants." Then she shifted intentionality to the programmer, and viewed the programmer on a psychological level. Finally she shifted to a mechanistic explanation, in which the programmer explicitly told the machine what to do. Which is the correct level? That is a natural but misleading question. Complex systems can be meaningfully described at many different levels. Which level is "best" depends on the context: on what you already understand and on what you hope to learn. In certain situations, for certain questions, the mechanistic level is the best. In other situations, for other questions, the psychological level is best. By playing with LEGO/Logo creatures, students learn to shift among levels, learning which levels are best for which situations. More generally, they learn how useful and powerful it is to think about systems in terms of levels.

Although LEGO/Logo can help students learn certain ideas about decentralized systems, it is not the ideal environment for exploring such systems. Imagine trying to use LEGO/Logo to re-create the behaviors of ants in an ant colony. You would need at least dozens, if not hundreds, of interacting LEGO/Logo robots. With our technology it is possible to create situations with two or three interacting robots, but certainly not dozens. Or what if you wanted to study evolution and natural selection? Building LEGO/Logo creatures that "give birth" to other LEGO/Logo creatures is far beyond the capabilities of our current technology.

To explore phenomena like these, I decided to go back to the computer screen, to work with virtual creatures in virtual worlds. Virtual worlds offer many advantages. In virtual worlds it is easy to create large numbers of creatures. It is easy to give new sensory capabilities to the creatures. And it is easy to set up and control precise experimental conditions.

Of course, there are some drawbacks to retreating to the computer screen. Watching a virtual creature wander to the edge of the computer screen doesn't have the same emotional impact as watching a LEGO/Logo creature bump into a wall. And certain aspects of the real world are very difficult to simulate accurately in virtual worlds. But that didn't bother me. I was not interested in perfect reproductions of the real world. Rather, I was interested in helping people explore the workings of decentralized systems—regardless of whether those systems are in the world or on the screen. In the next section I describe StarLogo, the programming language I developed to help people do that.

StarLogo

Using a computer to create and explore decentralized systems is certainly not a new idea. Over the years, computer scientists have developed a wide variety of decentralized computational models—such as neural networks, the subsumption architecture, and cellular automata. In all of these models, orderly patterns can arise from interactions among a decentralized collection of computational objects. In neural networks, patterns of "activation" arise from interactions among low-level "nodes." With the subsumption architecture (Brooks 1991) actions of a robotic creature arise from interactions among low-level "behaviors."

Cellular automata were probably the most influential on my thinking. In cellular automata, a virtual world is divided into a grid of "cells." Each cell holds a certain amount of "state." (On the computer screen, different states are usually represented by different colors.) In the simplest cases, each cell might hold just a single piece of state, indicating whether the cell is "alive" or "dead." There is a transition rule that determines how each cell changes from one generation to the next. Transition rules are typically based on the states of a cell's "neighbors." For example, a cell might become "alive" if the majority of its neighboring cells are alive. Each cell executes the same rule, over and over.

Cellular automata have proved to be an extraordinarily rich framework for exploring self-organizing phenomena. Simple rules for each cell sometimes lead to complex and unexpected large-scale structures. During the past two decades computer scientists and hackers have spent millions of computer-hours playing with various versions of cellular automata (including Conway's Game of Life, the best-known version). Through this experimentation they have developed a diverse set of examples and a rich language for describing cellular automata.

But cellular automata seem best suited as a tool for computer aficionados, not for the masses. The idea of writing "transition rules" for "cells" is not an idea that most people can relate to. Other decentralized computational models (such as neural nets) have similar drawbacks. In most cases, the objects being programmed (such as nodes and cells) seem too low level. These models are a great way for computer hackers and mathematicians to explore decentralized phenomena, but they seem ill suited for people who have less experience (or less interest in) manipulating formal systems.

I wanted to create a system in which the objects were more familiar, more related to people's experiences. Rather than exploring how the actions of individual creatures arise from interactions among "nodes" or "behaviors" or "cells," I decided to focus on how colony-level behaviors arise from interactions among individual creatures—for example, how ant-colony behaviors arise from interactions among individual ants. People are quite familiar with both of these "levels" (the creature level and the colony level). So I expected that people would be more interested in (and have a better chance at understanding) situations involving these levels. (I am using the terms "creature" and "colony" rather broadly. On a highway, each car can be considered a "creature," and a traffic jam can be considered the "colony.")

Logo seemed like a good starting point for my computational system. I wanted the system to encourage constructionist activities. I didn't want users merely to manipulate parameters in a standardized application program; I wanted them to construct and modify programs on their own, exploring situations of interest to them. Logo is well designed for such constructionist activities. The Logo turtle can be used to represent almost any type of object in the world: an ant in a colony, a car in a traffic jam, an antibody in the immune system, or a molecule in a gas. In addition, Logo has proved to be a relatively easy language for nonexpert programmers to learn and use.

On the other hand, traditional versions of the Logo language are missing many features that are needed for explorations of colony-type behaviors. So I set out to develop a new, extended version of Logo. This new version of Logo, which I call StarLogo, extends Logo in three major ways.

First, *StarLogo has lots more turtles.* Whereas commercial versions of Logo typically have only a few turtles, StarLogo has *thousands* of turtles. And StarLogo is designed as a *massively parallel* language—so all of the turtles can perform their actions at the same time, in parallel. For many colony-type explorations, having lots of turtles is not just a nicety, it is a necessity. In many cases the behavior of a colony changes qualitatively when the number of turtles is increased. An ant colony with ten ants might not be able to make a stable pheromone trail to a food source, whereas a colony with a hundred ants might. Similarly, a group of ten slime-mold cells might not aggregate into a cluster, whereas a hundred cells (in the same space, following the exact same rules) might.

Second, *StarLogo turtles have better "senses."* The traditional Logo turtle was designed primarily as a "drawing turtle," for creating geometric shapes and exploring geometric ideas. But the StarLogo turtle is more of a "behavioral turtle." StarLogo turtles come equipped with "senses." They can detect (and distinguish) other turtles nearby, and they can "sniff" scents in the world. There is even a built-in primitive to make turtles "follow the gradient" of a scent—that is, to make turtles turn in the direction where the scent is strongest. Such turtle-turtle and turtle-world interactions are essential for creating and experimenting with self-organizing phenomena. Parallelism alone is not enough. If each turtle just acts on its own, without any interactions, interesting colony-level behaviors will never arise.

Third, *StarLogo reifies the turtles' world.* In traditional versions of Logo the turtles' world does not have many distinguishing features. The world is simply a place where the turtles draw with their pens. Each pixel of the world has a single piece of state information—its color. StarLogo attaches a much higher status to the turtles' world. The world is divided into small square sections called *patches.* (The term *patch* is borrowed from Pauline Hogeweg [1989].) The patches have many of the same capabilities as turtles—except that they cannot move. Each patch can hold an arbitrary variety of information. For example, if the turtles are programmed to release a "chemical" as they move, each patch can keep track of the amount of chemical that has been released within its borders. Each patch might also keep track of how much "food" exists within its borders.

Patches can execute StarLogo commands, just as turtles do. For example, each patch could diffuse some of its "chemical" into neighboring patches, or it could grow "food" based on the level of chemical within its borders. Thus the environment is given equal status to the creatures that inhabit it. Other "creature-oriented" programming languages, such as Luc Steels's RDL (Steels 1989), tend to treat the environment as a passive entity, manipulated by the creatures that move within it. This view, not surprisingly, matches the way many people view the Earth itself. By reifying the environment, StarLogo aims to change the way people think about creature-environment interactions—perhaps leading to new and richer ways of thinking about how phenomena emerge in the world.

For example, the existence of patches encourages new ways of thinking about communications among creatures. Instead of communicating with one another directly, StarLogo creatures can communicate indirectly through the environment, by releasing chemical pheromones into the patches. This approach resembles the way that many social insects communicate with one another. Similarly, creatures can leave "reminder markers" in the environment instead of burdening their own memories. This idea of making use of objects in the environment, rather than creating new internal representations, is an example of what is sometimes known as "distributed cognition."

StarLogo patches are much like the cells in cellular automata. Thus StarLogo programs can often be conceptualized as turtles moving on top of (and interacting with) a cellular-automata grid. All types of interactions are possible: turtle-turtle interactions, turtle-patch interactions, patch-patch interactions. StarLogo includes several special-purpose primitives (such as `follow-gradient`) to facilitate interactions between turtles and patches, and others (such as `diffuse`) to facilitate interactions between neighboring patches. StarLogo places special emphasis on *local* interactions—that is, interactions among turtles and patches that are spatially near one another. Thus the language is well suited for explorations of self-organizing phenomena, in which large-scale patterns arise from local interactions.

In some ways the ideas underlying StarLogo parallel the ideas underlying the early versions of Logo itself. In the late 1960s Logo aimed to make then-new ideas from the computer science community (such as procedural abstraction and recursion) accessible to a larger number of users. Similarly, StarLogo aims to make 1990s ideas from computer science (such as massive parallelism) accessible to a larger audience. And whereas Logo introduced a new object (the turtle) to facilitate explorations of particular mathematical/scientific ideas (such as differential geometry), StarLogo introduces another new object (the patch) to facilitate explorations of other mathematical/scientific ideas (such as self-organization).

StarLogo and Stella

StarLogo shares some common goals with Stella, a modeling tool that grew out of research in the field of system dynamics (for example, Forrester 1971; Roberts et al. 1983). Both StarLogo and Stella are designed to help people explore the behaviors of systems. But there are some important differences. In using Stella, you typically think in terms of aggregate quantities. For example, if you want to model an ecosystem with rabbits and foxes, you specify the factors that affect the overall population of rabbits and the factors that affect the overall population of foxes. Then Stella generates a graph showing how the rabbit and fox populations evolve with time. StarLogo, by contrast, focuses more on individual creatures, not overall populations. To model the same ecosystem, you specify how each individual rabbit behaves and how each individual fox behaves. Then StarLogo displays the creatures running around, and interacting with one another, on the screen. The behavior of the overall population arises out of the interactions among the individual creatures.

Sample Session with StarLogo

Probably the best way to learn about StarLogo is to sit down at a computer and play with it. This section tries to re-create that experience, presenting a sample session with StarLogo (see appendix B for more information about StarLogo).

```
create-turtle 100
```
100 turtles appear on the screen. By default, the turtles start with random positions and random headings.

`forward 200`
All turtles move forward 200 "turtle steps." The turtles "wrap" at the edges of the screen. That is, when turtles go off the right edge, they wrap around to the left edge. And when they go off the top edge, they wrap around to the bottom edge. The turtles do not draw as they move: unlike traditional Logo turtles, StarLogo turtles start with their "pens" up.

`forward random 200`
Each turtle chooses a different random number (between 0 and 199), so each moves forward a different distance.

`set-color blue`
The turtles turn blue.

`if ypos < 0 [set-color green]`
The position (0,0) is at the center of the screen. So only turtles in the bottom half of the screen have *y*-positions (`ypos`) less than 0. Those turtles turn green.

`pendown forward 200`
Like standard Logo turtles, StarLogo turtles have "pens" for drawing. A turtle's pen is always the same color as the turtle itself. In this case, each turtle draws a straight line, 200 turtle steps long.

`if color = green [repeat 36 [forward 1 right 10]]`
Each green turtle draws a circle.

`clear-patches`
Clears the color from the "patches" that make up the turtles' world. This command plays the same role as `clear-screen` or `clear-graphics` in traditional versions of Logo.

`turtles-have step-size`
Each turtle creates a variable named step-size. (In computer science terms, `step-size` is a "local state variable" for each turtle. Turtles begin with a small core of local state variables (position, heading, color), but users can add arbitrary new state variables.)

`set-step-size random 100`
Each turtle sets its value for `step-size` to a random number between 0 and 99. Each turtle chooses its own random number, so the turtles will (generally) have different values for `step-size`. (Note: When the variable `step-size` was created, StarLogo automatically created two new procedures: the procedure `set-step-size` to change the value of the variable, and the procedure `step-size` to return the value of the variable.)

```
forward step-size
```
The turtles move forward different distances, depending on their values for step-size.

```
clear-all
```
Clears the patches and "kills" all of the turtles.

```
set-patchcolor yellow
```
Each patch sets its color to yellow. So the whole background of the screen turns yellow.

```
if patch-xpos > 20 [set-patchcolor green]
```
Each patch with an *x*-position greater than 20 turns green.

```
if (distance 10 20) < 15 [set-patchcolor white]
```
Each patch checks to see if its distance from the point (10, 20) is less than 15 units. If so, it turns white. The result: a white disk of radius 15, centered on the point (10, 20).

```
clear-patches
```
Clears the color from all of the patches.

```
patches-have chemical
```
Each patch creates a local state variable named chemical.

```
set-chemical 0
```
Each patch sets the value for chemical to 0. (Note that each patch could have a different value for chemical. But for now, the value for chemical is the same (zero) in every patch.)

```
if (distance 0 0) < 20 [set-chemical 30]
```
Patches near the middle of the screen (inside a circle of radius 20) set the value of chemical to 30. In other patches, the value of chemical remains 0. Nothing changes on the display, since the command didn't tell the patches to change color.

```
diffuse chemical
```
Each patch "diffuses" its chemical—that is, it spreads its value for chemical evenly among its eight neighbors. As a result, the "chemical puddle" in the middle of the screen spreads outward a little bit. (But still, nothing changes on the display, since the diffuse command does not make the patches change color.)

```
repeat 50 [diffuse chemical]
```
The patches diffuse the chemical for 50 more iterations, making the "chemical puddle" spread further outward.

```
scale-patchcolor green chemical 0 5
```
The `scale-patchcolor` primitive uses varying intensities of a single color to represent the varying values of a variable. In this case, it uses varying intensities of green to represent varying values of `chemical`. Patches near the center of the screen (where the chemical is most intense) turn bright green, while patches further from the center are dimmer shades of green. (Patches in which the value of `chemical` is 5 (or more) set their color to the brightest green; patches in which the value of `chemical` is 0 (or less) set their color to "no-intensity green" (black); patches in which the value of `chemical` is between 0 and 5 set their color to some intermediate intensity of green.)

```
repeat 100 [diffuse chemical
            scale-patchcolor green chemical 0 5]
```
Continues the diffusion for 100 more iterations, updating the display on every step.

```
create-turtle 200
```
Creates 200 turtles.

```
repeat 100 [set-heading uphill chemical forward 1]
```
`uphill` is used to "follow a gradient" of a variable. In this case each turtle "sniffs" in several directions and sets its heading in the direction where the value of `chemical` is largest. Before each step the turtles set their headings to line up with the chemical gradient. After 100 iterations all turtles end up at the center of the screen, where the value of `chemical` is largest.

Procedures and Demons

It is useful to think of StarLogo procedures in three categories. Some procedures are meant to be executed by turtles. Other procedures are meant to be executed by patches. Still other procedures are meant to be executed by an entity that I call the StarLogo *observer*. In a sense, the StarLogo observer looks down on the turtles and patches—much like a god looking down on the creatures and environment below. The observer is very useful for setting up initial conditions and for monitoring the overall activity in the StarLogo world. For example, the observer can calculate the number of turtles on the right side of the screen (with the command `tur-tle-subtotal [xpos > 0]`) or the total amount of food in all of the patches (with the command `patch-sum [food]`).

A sample StarLogo program follows. The `setup` procedure creates two hundred new turtles, then executes the `patch-setup` procedure—

which, in turn, creates a circular "puddle" of blue chemical at the center of the screen. The walk procedure makes the turtles wander randomly around the screen. If a turtle happens to walk over the blue chemical, it turns yellow (as if in reaction to the chemical).

Observer Procedures

```
to setup
clear-all
create-turtle 200
patch-setup
end
```

Turtle Procedures

```
to walk
random-step
test-for-chemical
end

to random-step
right random 40
left random 40
forward 1
end

to test-for-chemical
if any-chemical-here?
   [set-color yellow]
end

to any-chemical-here?
ask patch-here [chemical > 0]
end
```

Patch Procedures

```
to patch-setup
patches-have chemical
ifelse (distance 0 0) < 20
   [set-chemical 10 set-patchcolor blue]
   [set-chemical 0]
end
```

To run this StarLogo program, you would type something like this:

```
setup
repeat 1000 [walk]
```

As the turtles walk around the screen, many of them (by chance) wander through the chemical puddle and turn yellow. If you execute the second command again, even more of the turtles will wander through the puddle and turn yellow. If you execute the command enough times, eventually all of the turtles will turn yellow.

An alternate approach for running this program is to use StarLogo's *demon* mechanism (Evans 1991). If you activate a procedure as a demon, the procedure keeps executing over and over. For example, you can activate the walk procedure as a demon:

```
activate-demon walk
```

As soon as you execute this command, the walk procedure begins to execute repeatedly. As in the previous approach, each turtle walks around the screen, turning yellow if it happens to wander into the chemical puddle. The walk procedure continues to execute until you explicitly deactivate the demon (or tell StarLogo to stop all demons).

The demon approach offers some major advantages, since demons execute in the "background." That means you can type other commands *while* the demons are running. So while the turtles are wandering around the screen, you can execute this command:

```
turtle-subtotal [color = yellow]
```

StarLogo immediately prints out the number of yellow turtles, without any apparent interruption in the execution of the turtles' walk procedure. If you execute the same command a few seconds later, you might get a different result, since more of the turtles might have wandered through the chemical puddle. Or you can add a new chemical puddle by executing this command:

```
if (distance 30 40) < 10
        [set-chemical 10 set-patchcolor blue]
```

A new chemical puddle appears immediately, and the turtles turn yellow when they wander through the new puddle, just as they do with the original puddle.

In this example, only one procedure (walk) is activated as a demon. But you can activate as many demons as you would like, and they will all keep executing over and over. For example, instead of activating the walk procedure as a demon, you could activate its two subprocedures (ran-

dom-step and test-for-chemical) as demons. Then StarLogo
would repeatedly execute *both* of these procedures (interweaving the exe-
cutions in an unspecified way). Of course, this two-demon approach
would not work if the program depended critically on the order of execution of the two subprocedures. But in this case the ordering is not
important. So the decision to activate the two subprocedures (rather than
the walk procedure) actually serves an important role: it highlights the
fact that the order of execution of the subprocedures is unimportant.

Implementation of StarLogo

The first version of StarLogo was implemented on the Connection
Machine, a massively parallel computer with thousands of processors
(Hillis 1985). The name *StarLogo* follows the convention for other
Connection Machine languages: the version of Lisp for the Connection
Machine is called *Lisp (pronounced star-lisp), and the version of C is
called C* (pronounced C-star). I implemented StarLogo in *Lisp.

Massively parallel hardware is not theoretically important to the
StarLogo approach. StarLogo aims to provide the programmer with a
clear *conceptual model* for massive parallelism. The programmer writes programs *as if* the creatures and patches are all acting in parallel, regardless of
the underlying implementation. It was very convenient for me to implement the first version of StarLogo on massively parallel hardware, since I
was able to implement StarLogo in a high-level language, without worry-
ing about speed. But StarLogo has been reimplemented (in lower-level
code) on several sequential machines (including UNIX machines and the
Macintosh), to make the language accessible to more people.

Objects and Parallelism

During the past decade, many ideas have influenced the design of computer programming languages. But two ideas stand out as especially influential: object-oriented programming and parallelism. StarLogo incorpo-
rates aspects of both object-oriented programming and parallelism, but it
is a bit out of the mainstream in its approach to these ideas. Its distinctive
approach is due, in part, to its distinctive goals. Designers of most programming languages are interested in improving speed and performance,
or in providing advanced capabilities for expert programmers. StarLogo,
by contrast, is designed for a larger group of users, including nonexpert
programmers. And it focuses not on improving the speed of computation

but on helping people think about and experiment with important scientific ideas in new ways.

This section discusses StarLogo's approach to object-oriented programming and parallelism, analyzing how the design choices in StarLogo reflect the underlying goals of the language.

Object-Oriented Programming

In recent years there has been a great deal of enthusiasm over object-oriented programming. In professional programming circles the appeal of object-oriented programming is clear from the proliferation of languages in which object-oriented features are grafted onto traditional languages such as Lisp or C. Among nonexpert programmers, object-oriented programming has received wide exposure through the dissemination of HyperCard and its underlying programming language HyperTalk (which is based on a set of special objects including buttons, cards, and stacks). The success of HyperCard shows the power of even limited forms of object-oriented computation, and it draws attention to the appropriability of object-oriented programming by novices.

The excitement over object-oriented programming has spread far beyond computer programmers and researchers. In 1991 *Business Week* magazine devoted an entire cover story to object-oriented programming (under the title "Software Made Simple"), arguing that object-oriented approaches would greatly simplify the task (and thus reduce the cost) of producing computer software. Clearly, some of these claims are greatly exaggerated. Major software projects will not become "simple." Nevertheless, certain object-oriented features have the potential to make some computer programs easier to create, maintain, extend, and understand.

In some ways Logo was one of the first object-oriented languages. The Logo turtle, after all, is a type of object. The turtle comes with a collection of local state—color, heading, shape, pen state. And turtle commands (like forward, back, left, and right) can be viewed as messages that the turtle objects understand. But traditional versions of Logo offer a very limited form of object-oriented programming. Other programming languages (such as Smalltalk and C++) offer many more object-oriented features. For example, users can create arbitrary new "classes" of objects, and establish "inheritance hierarchies" to share properties among those classes. Does it make sense to add these features to Logo? If so, which object-oriented features are most important?

There is already one commercial version of Logo, called Object Logo, that includes a rich collection of object-oriented features. Object Logo is beautifully designed from a computer science point of view. It offers almost all of the object-oriented features that anyone could imagine, and it integrates them into Logo in an elegant fashion. But are all of these object-oriented features necessary, or even desirable, for the nonexpert programmer?

Many object-oriented features are designed to support modularization of programs. They allow programmers to conceptualize and implement programs as a collection of objects that "connect" with one another in standardized ways—much like a collection of LEGO bricks. In the ideal case, each type of object can be tested, modified, and extended on its own, without affecting any of the other objects in the system.

Such modularization can be very valuable in developing complex programs. But for nonexpert programmers, the object-oriented approach is valuable for a different set of reasons. When programs are based on objects, programs often become easier for nonexperts to relate to and think about. Telling a turtle to take 50 steps forward has a very different "feel" from telling the computer to draw a line between the Cartesian coordinates (23, 57) and (71, −14). Students can relate to the turtle, even imagine themselves as the turtle. In the words of Seymour Papert, the turtle becomes an "object to think with."

In education and cognitive science research, there is a growing emphasis on so-called concrete and situated approaches to learning. Educational researchers are starting to steer away from the formal, logical approaches that have dominated mathematics and science education in the past, moving toward more contextualized approaches involving objects and situations familiar to the students. This shift is part of a broader intellectual trend that Turkle and Papert (1990) have dubbed a "revaluation of the concrete." Object-oriented programming fits firmly in this trend, encouraging people to think about programming in terms of concrete objects (like turtles and buttons) rather than propositional rules. (As Turkle and Papert point out, it is ironic that the computer, commonly viewed as the ultimate embodiment of logical, abstract thinking, should lend support to a revaluation of the concrete.)

So as I designed object-oriented features into StarLogo, I was not terribly concerned with advanced object-oriented features, such as inheritance hierarchies. I was more interested in providing users with good "objects to think with." I wanted objects that would encourage and support certain types of explorations (in particular, explorations of decentralized systems and self-organizing behaviors). I wanted the objects to be familiar enough for users to relate to, and general enough to support a diverse set of projects.

There is an old saying that goes something like this: If a person has only a hammer, the whole world looks like a nail. Indeed, a person's perceptions and models of the world are strongly shaped by the objects that exist in the world. The same is true for computational systems. The way people interact with (and think about) a computational system depends strongly on the objects that compose the system. If the objects are well chosen for the intended task, even novices will use the system productively. If the objects are not well chosen, even experts will struggle.

With these thoughts in mind, I based StarLogo on two primary types of objects: turtles and patches. The Logo turtle is a simple yet "high-leverage" object: it has inspired thousands of diverse projects, explorations, and experiments. And the Logo turtle is particularly well suited for explorations of self-organizing behaviors. The design of the Logo turtle emphasizes "local" over "global"—turtle geometry programs use local coordinates rather than Cartesian coordinates. Explorations of self-organizing behaviors have a similar emphasis, relying on local interactions to produce large-scale patterns.

Perhaps more important, the Logo turtle (unlike the cells in cellular automata) is an object that students can easily relate to, and even identify with. One high-school student used two colors of StarLogo turtles (red and yellow) to represent two different species of creatures in an ecosystem. She explained the importance of the turtle/creature metaphor: "If you just said these red dots don't like the yellow dots, I wouldn't have been interested. It's the context that you put it in. If you describe anything just in terms of dots on the screen, it's not very interesting."

By adding the patch as a new computational object, I hoped to provide StarLogo users with another powerful object to think with. The StarLogo patch encourages people to adopt a new view of the "environment" in which creatures live. Too often, people view the environment as a passive entity. Science, literature, and popular culture have all reinforced this view. Commenting on the film *Dances with Wolves*, a Western film that won the Academy Award for best picture in 1990, historian Richard White noted, "Nature always serves as a setting in the Western, but this one makes nature a character" (quoted in Dumanoski 1991). StarLogo tries to do something similar. By bestowing "objecthood" on the environment (in the form of the patch), StarLogo encourages users to think of the environment as a character, not just a background setting.

Parallelism

During the past two decades there has been ever-increasing interest in parallel computation and parallel programming languages. Some researchers have added parallel constructs to existing programming languages, yielding new language dialects like Concurrent Pascal (Brinch Hansen 1975) and MultiLisp (Halstead 1985). Other researchers have created entirely new programming models, designed explicitly with parallelism in mind (for example, Sabot 1988).

In most of this research the primary goal is to improve the speed of computation. A recent article in a major computer science journal quotes a user saying, "Nobody wants parallelism. What we want is performance" (Pancake 1991). In other words, many people see parallelism as a necessary evil in order to improve the speed at which programs execute. If they could, many language developers would hide parallelism from the user. Indeed, some researchers have developed parallelizing compilers, which allow programmers to continue writing programs in traditional sequential style, putting the burden on the compiler to "parallelize" the code to improve performance.

In adding parallelism to Logo, I had a very different set of goals. I was not particularly concerned with performance or speed. Rather, I was interested in providing easier ways for programmers to model, simulate, and control actions that actually happen in parallel. Many things in the natural world (such as ants in a colony) and the manufactured world (such as rides in an amusement park) really do act in parallel. The most natural way to model and control such situations is with a parallel programming language. In these cases parallelism isn't a trick to improve performance; it is the most natural way of expressing the desired behavior.

Before beginning to work on StarLogo, I had designed another parallel version of Logo, called MultiLogo (Resnick 1990). The design of MultiLogo was motivated by my research on LEGO/Logo. When children build and program LEGO machines, they often want different machines to run different programs at the same time. For example, after building an amusement park with a LEGO Ferris wheel and a LEGO merry-go-round, a child might want the two rides to run different programs at the same time. That is a very natural thing to want. In fact, many children are surprised that traditional versions of Logo can't do such a simple thing. Running simultaneous programs is a simple thing to think about, shouldn't it be a simple thing to do in Logo? Indeed, the lack of parallelism in Logo violates the principle that simple things should be simple.

MultiLogo solved this problem, adding a simple way for users to create and execute multiple processes at the same time. MultiLogo adds one new programming construct: the agent. Each agent is like a separate version of Logo—that is, each agent can control a computational process. By using multiple agents, users can control multiple processes. In this way, MultiLogo users can control the simultaneous actions of multiple LEGO machines in the world, or multiple Logo turtles on the screen. (There is now a commercial version of Logo, called MicroWorlds, with similar multiprocessing features.)

StarLogo is designed with different sorts of situations in mind. Whereas MultiLogo is designed for situations with a few parallel processes, StarLogo is designed for situations with hundreds or thousands of objects acting in parallel (such as ants in an ant colony). And whereas the objects in MultiLogo often behave quite differently from one another, objects in StarLogo are usually quite similar to one another. In general, each StarLogo turtle does roughly the same thing as each of the others, and each patch does roughly the same thing as each of the others.

For me, situations with lots of nearly identical objects have a special aesthetic and intellectual appeal. It is more intriguing if a complex, orderly pattern arises from interactions among simple, homogeneous objects than if the same pattern arose from interactions among complex, heterogeneous objects. StarLogo is explicitly designed to help people explore such complexity-from-simplicity situations.

From One to Many

In the past, Logo typically meant one student sitting in front of one computer using one process to control one turtle in one medium (graphics). Designers of future versions of Logo should aim to change *one* to *many*. In the future, Logo should support many users (communicating over networks), many objects (thousands of turtles, for example), many types of objects (not just turtles, but patches and buttons and other things), many processes (process-level parallelism), and many media (video and music and robotics in addition to graphics).

StarLogo starts to move in this direction. Its most obvious contribution is its support for thousands of turtles. It also introduces a new type of object (patches), and a limited form of multiple processes (demons). But StarLogo is just a beginning. There are many more things to many-ize.

Looking Ahead: From Constructions to Explorations

There is an old proverb that states, "Hear and forget; see and remember; do and understand." This proverb represents a rather simplistic theory of learning. But, like all good proverbs, it hints at a deeper and more complex truth. Many of our richest learning experiences grow out of situations in which we are engaged in designing and constructing personally meaningful things.

StarLogo brings this constructionist approach to the study of decentralized systems: it allows people to construct and play with decentralized systems. Next, in Explorations, I describe nine decentralized systems that high-school students and I constructed and explored with StarLogo.

3

Explorations

Go to the ant, thou sluggard;
consider her ways, and be wise.

—The Bible (Proverbs 6.6)

Simulations and Stimulations

I'm not quite sure what to call the type of projects that I've worked on
with StarLogo. Most people would call them *simulations*. But that doesn't
feel quite right. A simulation is, according to *Webster's New Collegiate
Dictionary*, "the imitative representation of the functioning of one system
or process by means of the functioning of another." Many computer simulations fit this description. They try to imitate some real-world system
or process as accurately as possible. In many cases computer simulations
are used to make predictions about real-world processes. Computer simulations of nuclear reactors are used to predict when the reactors might
fail. Computer simulations of meteorological patterns are used to predict
tomorrow's weather. In these cases, the more accurate the simulation, the
better.

In working with StarLogo, I have different goals. To be sure, many
StarLogo projects are inspired by real-world systems: ant colonies, forest
fires, traffic jams. But I'm not interested in developing accurate imita-
tions of these real-world systems, or even in making accurate predictions
about them. The real world serves only as an inspiration, a departure
point for thinking about decentralized systems. When I write a StarLogo
program with artificial ants, for instance, I am more interested in investi-
gating antlike behaviors than the behaviors of real ants. Even more, I am
interested in how people *think about* antlike behaviors.

In short, I am more interested in *stimulation* than in *simulation*. My
work with StarLogo is aimed more at what's *in here* (in the mind) than

what's *out there* (in the world). The goal is not to simulate particular systems and processes in the world. The goal is to probe, challenge, and disrupt the way people think about systems and processes in general.

I prefer to think of StarLogo projects as *explorations of microworlds*, not simulations of reality. Microworlds are simplified worlds, specially designed to highlight (and make accessible) particular concepts and particular ways of thinking. Microworlds are always manipulable: they encourage users to explore, experiment, invent, and revise. Seymour Papert (1980) describes microworlds as "incubators for knowledge." The standard Logo-turtle microworld, he writes, is a place where "certain kinds of mathematical thinking [can] hatch and grow with particular ease." StarLogo, with its sensory-enhanced turtles and patch-reified environment, is particularly well suited for developing "systems science" microworlds—that is, worlds where systems thinking can hatch and grow.

In the rest of this chapter I describe and discuss explorations of some StarLogo microworlds. In some of the explorations I worked by myself. In others I worked with students from Boston-area high schools. (For more information on the high-school students, see appendix A.)

When I worked with high-school students, I played several (simultaneous and intertwined) roles:

Observer. I observed how the students thought about decentralized systems, and how their thinking evolved during their interactions with StarLogo.

Catalyst. I proposed experiments, asked questions, challenged assumptions, and encouraged students to reflect on their experiences as they worked with StarLogo.

Collaborator. I helped students write their StarLogo programs, since I was not particularly interested in studying how well students learned to program in StarLogo. More important, I worked together with students in trying to make sense of unfamiliar phenomena. Often, working with students helped clarify my own thinking about decentralized systems.

Slime Mold

Slime mold is hardly the most glamorous of creatures. But it is surely one of the most strange and intriguing. As long as food is plentiful, slime-mold cells exist independently as tiny amoebas. They move around, feed on bacteria in the environment, and reproduce simply by dividing into two. But when food becomes scarce, the slime-mold behavior changes dramatically. The slime-mold cells stop reproducing and move toward one another, forming a cluster (called a "pseudoplas-modium") with tens of thousands of cells.

At this point, the slime-mold cells start acting as a unified whole. Rather than behaving like lots of unicellular creatures, they act as a single multicellular creature. In short, "they" start acting like "it." It changes shape and begins crawling, seeking a more favorable environment. When it finds a spot to its liking, it differentiates into a stalk supporting a round mass of spores. These spores ultimately detach and spread throughout the new environment, starting a new cycle as a collection of slime-mold amoebas.

The process through which slime-mold cells aggregate into a single multicellular creature has been a subject of scientific debate. Until 1980 or so, most biologists believed that specialized "pacemaker" cells coordinated the aggregation. But scientists now view slime-mold aggregation as a very decentralized process. According to the current theories, slime-mold cells are homogeneous: none is distinguished by any special features or behaviors. The clustering of slime-mold cells arises not from the commands of a leader but through local interactions among thousands of identical cells. In fact, the process of slime-mold aggregation is now viewed as one of the classic examples of self-organizing behavior.

How do the slime-mold cells aggregate? The mechanism involves a chemical called "cyclic AMP" or cAMP (Goldbeter and Segal 1977). When the slime-mold cells move into their aggregation phase, they pro-

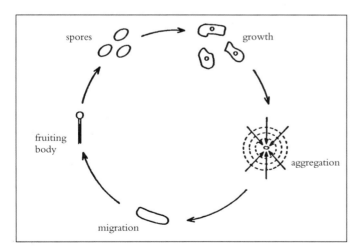

Figure 3.1
Life cycle of slime mold, reproduced from Prigogine and Stengers 1984

duce and emit cAMP into the environment. They are also attracted to the very same chemical. As the cells move, they follow the gradient of cAMP. That is, they test around themselves, and they move in the direction where the concentration of cAMP is highest. Note that this process is very *local*. Each cell can sense cAMP only in its immediate vicinity; it cannot tell how much cAMP there might be a few centimeters away.

I wrote a StarLogo program to explore the workings of this decentralized aggregation process. I was not interested in simulating every detail of the actual slime-mold mechanism. In the actual mechanism, slime-mold cells produce the cAMP in periodic pulses. As a result, slime-mold cells tend to come together in concentric waves. But this periodicity does not seem essential to the aggregation process. In fact, Prigogine and Stengers (1984) describe how the larvae of certain beetles (*Dendroctonus micans*) aggregate into clusters using a mechanism similar to that used by slime-mold cells, but without the periodicity.

My goal was to capture the essence of the aggregation process with the simplest mechanism possible. My StarLogo program was based on a set of simple rules. Each turtle was controlled by four demons: one demon made the turtle move, a second added a little randomness to the turtle's movements, a third made the turtle emit a chemical pheromone, and a fourth made the turtle "sniff" for the pheromone and turn in the direction where the chemical was strongest (that is, follow the gradient of the pheromone).

Meanwhile, each patch was controlled by two primary demons: one to make the pheromone in the patch evaporate, and another to diffuse the pheromone to neighboring patches. (A third demon controlled the color of the patches. Each patch was displayed as a shade of green: the more pheromone in the patch, the brighter the intensity of green.) All of the demons (for the turtles and patches) were very simple; each required at most two lines of StarLogo code.

Observer Procedures

```
to setup number
clear-all
create-turtle number
turtle-setup
patch-setup
end
```

Turtle Procedures

```
to turtle-setup
activate-demon [walk-demon wiggle-demon
               drop-pheromone-demon sniff-demon]
end

to walk-demon
forward 1
end

to wiggle-demon
right random 40
left random 40
end

to drop-pheromone-demon
ask patch-here [set-pheromone pheromone + 1]
end

to sniff-demon
if ask patch-here [pheromone > 2]
   [follow-gradient pheromone]
end
```

Patch Procedures

```
to patch-setup
patches-have pheromone
activate-demon [evaporation-demon
               diffusion-demon display-demon]
end

to evaporation-demon
set-pheromone pheromone * 0.9
end

to diffusion-demon
diffuse pheromone
end

to display-demon
scale-patchcolor green pheromone 0 3
end
```

Programming Notes

On each iteration of the program the demon procedures are executed in an unspecified order. In my example programs I tend to use multiple demons (instead of multiple subprocedures within a single procedure) when the order of execution is unimportant. In this program the order of execution does affect the detail of what happens. But the overall effect of aggregation remains the same, regardless of the order in which the demons are executed.

By convention, I usually append -demon to the names of procedures that are used as demons.

If we start the simulation with a small number of turtles, not much happens. We see faint green trails of pheromone behind each turtle. But these trails quickly dim as the pheromone evaporates and diffuses. Sometimes a turtle will follow another turtle for a short while, but it quickly loses the trail. Overall, the screen has a faint green aura, indicating a low level of pheromone everywhere, but no bright green areas. The turtles seem to wander aimlessly, looking somewhat like molecules in a gas.

But if we add enough turtles to the simulation, the behavior changes dramatically. With lots of turtles, there is a better chance that a few turtles will wander near one another. When that happens, the turtles collectively drop a fair amount of pheromone, creating a sort of pheromone "puddle" (shown as a bright green blob on the display). The turtles in the puddle, by following the pheromone gradient, are likely to stay within the puddle—and drop even more pheromone there, making the puddle even bigger and more "powerful." And as the puddle expands, more turtles are likely to "sense" it and seek it out—and drop even more pheromone (figure 3.2).

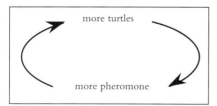

Figure 3.2
Positive feedback

The result is a self-reinforcing positive feedback loop: (1) the more pheromone in the puddle, the more turtles it attracts, and (2) the more turtles attracted to the puddle, the more pheromone they drop in the puddle.

With enough turtles, this same process can play out in many locations, resulting in turtle/pheromone clusters all over the computer screen. Through the positive feedback mechanism, the clusters tend to grow larger and larger (figure 3.3). What's to stop the clusters from growing forever? The positive feedback loop is balanced by a negative

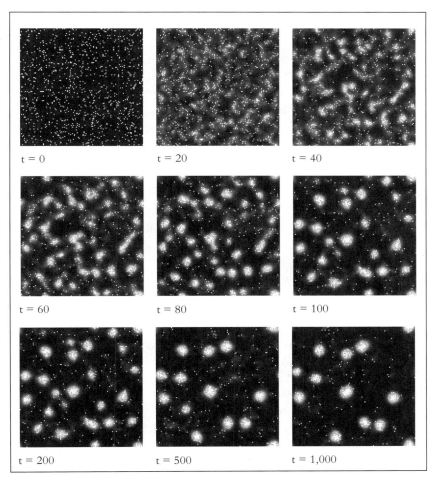

t = 0   t = 20   t = 40

t = 60   t = 80   t = 100

t = 200   t = 500   t = 1,000

Figure 3.3
1,000 iterations with 1,000 slime-mold cells

feedback process: as the clusters become bigger, there are fewer free turtles wandering around the world, depriving the positive feedback process of one of the raw materials that it needs to keep going. For the clusters to keep growing, the system would need a never-ending supply of new turtles.

As the turtles wander around the screen, there is a bit of randomness in their motion (from the `wiggle-demon` procedure). This randomness serves one obvious purpose: it ensures that free turtles will eventually wander near some cluster. Once a free turtle wanders near a cluster, it senses the pheromone from the cluster and begins to follow the gradient of the pheromone. At that point the randomness might seem to play a negative role. Why would we want to cripple a turtle's ability to follow the pheromone?

In fact, the program would be quite boring if the turtles followed the pheromone perfectly. Eventually each turtle would join a cluster. After that, not much more would happen. Individual clusters could never grow larger or smaller, and the number of clusters would never change. Although turtles would still move around within their clusters, the composition of each cluster would be fixed. Turtles would never leave their clusters. The screen would be filled with stable, unchanging green blobs (with a little activity inside each blob).

A bit of randomness in the turtles' movements leads to a much more interesting dynamic. Turtles are not forever bound to the clusters they join. Sometimes, through its random motion, a turtle will break free of its cluster and begin wandering again. Such an escape can initiate a ripple effect. With one fewer turtle in the cluster, there is a little less pheromone in the cluster. So the cluster is a little less likely to attract new turtles, and a little more likely to lose some of its remaining turtles. If another turtle escapes, the cluster becomes even weaker, and even less likely to hold onto its remaining turtles. As a result, small clusters often break apart suddenly. One turtle escapes, and then another, and another, in rapid succession. Underlying this rapid disintegration is the same positive feedback process that drives the formation of clusters—but operating in the reverse direction.

In a similar way, nearby clusters tend to merge together. Imagine two neighboring clusters of roughly equal size. Call them cluster A and cluster B. Turtles are equally likely to escape from either cluster and join the other. But what if, by random chance, a few turtles happen to move from cluster A to cluster B? Then cluster A becomes weaker (with less pheromone), and more likely to lose even more turtles. Cluster B, meanwhile, becomes stronger (with more pheromone), and more likely to attract even more turtles. So the movement of turtles from cluster A to cluster B is likely to continue—and to accelerate.

So as the program proceeds, nearby clusters are likely to merge, and small clusters are likely to break apart, freeing turtles to join (and enlarge) the remaining clusters. As a result, the number of clusters tends to decline with time, and the number of turtles in each cluster tends to increase. As the clusters grow larger and larger, they become more and more stable. Turtles are less likely to escape. And even when an errant turtle escapes, it is less likely to set off a chain reaction destroying the entire cluster.

Will the turtles eventually join into a single, giant cluster? Everyone who sees the program seems to have that intuition. After all, the number of clusters declines with time, and the number of turtles in each cluster increases. Why shouldn't that trend continue to its natural conclusion: a single, giant cluster? This intuition is bolstered by the fact that a single, giant cluster would be very stable. If the turtles ever got together into a single, giant cluster, they would very likely remain together.

My guess is that the turtles would end up in a single, giant cluster—but only after a *very* long time. I ran the program with a thousand turtles, and left it running for several hours. In the first minute a couple dozen clusters formed. For the next few minutes the number of clusters declined, and the size of the remaining clusters increased. After a few minutes there were nine large clusters on the screen. But after that the system remained quite stable. After a couple of hours the number of clusters had declined by only one (to eight), the result of two clusters merging. Why the stability? Each of the nine clusters had reached a "critical mass," making them unlikely to break apart. The merger of two of the clusters was caused by "cluster drift"—the two clusters had drifted near one another, making it easier for turtles to jump from one to the other. But clusters drift very slowly. So when clusters are few and widely separated, mergers are not very likely. Only after very long periods of time are the clusters likely to join together.

The behavior of the slime-mold program varies significantly as parameters are changed. The program depends critically on the number of turtles. With too few turtles, no clusters form. If the density of turtles in the world rises above a certain critical density, clusters begin to form. With a higher density of turtles, there is a greater chance that a few turtles will wander near one another, forming a small pheromone puddle from which a cluster will grow. Of course, the exact critical density depends on many other factors. If the evaporation rate of the pheromone is increased, more turtles are needed to start a cluster (so the critical density is higher). If the turtles emit larger drops of pheromone on each step, fewer turtles are needed to start a cluster (so the critical density is lower).

What if we change the turtles' sense of smell? There are several ways to do that. One way is to change the range of directions that the turtles sniff. By default, each turtle takes three sniffs in trying to follow the gradient of a scent: one sniff straight ahead, one sniff 45 degrees to the left of its heading, one sniff 45 degrees to the right of its heading. (On each sniff the turtle senses one unit-distance away from its current position.) What if we make the turtles take more sniffs? Say each turtle takes five sniffs: 90 degrees to the left, 45 degrees to the left, straight ahead, 45 degrees to the right, and 90 degrees to the right. Equivalently, we could think of this as increasing the number of noses on each turtle, so that each turtle has five noses instead of three noses, equally spaced at 45-degree intervals. (We can make this change with the StarLogo command `set-number-of-sniffs 5`.) With five noses/sniffs rather than three, the turtles clearly have a better sense of smell. How will this improved sense of smell change the dynamics of the program? Will there be more clusters or fewer? Will the clusters be larger or smaller? Think about it a minute before reading on.

I posed this scenario to about two dozen people (including high-school students and MIT researchers). More than three-quarters of the people predicted the result incorrectly. Most people expected fewer and bigger clusters. In fact, the turtles gather into more and smaller clusters. It isn't too surprising that many people had difficulty predicting what would happen. After all, the slime-mold program involves thousands of interacting objects. It is very difficult to make predictions about such complex systems. So it wouldn't be too surprising if half of the people predicted the result incorrectly. But it seems strange that *most* people predicted incorrectly. What underlies this false intuition?

I asked people to explain their reasoning. Many people reasoned something like this: "The creatures are trying to get together, to combine into one big thing. If the creatures have a better sense of smell, they will do a better job of that. So you'll end up with larger clusters." What's the flaw? This reasoning confuses levels by attributing inappropriate intentionality to the creatures. Creatures are not really trying to form large clusters; they are simply following a pheromone gradient. The creatures *do* follow the gradient more effectively when they have more noses. But as a result they form smaller (not larger) clusters. By following the gradient effectively, the many-nosed creatures more quickly "find" other creatures to interact with. Giving more noses to the creatures is like giving a larger cross-section to particles in a physics simulation: collisions are more likely. And once the creatures find some others to interact with, they can form stable clusters with fewer partners, since each creature in the cluster stays closer to the others. The result is that clusters are smaller, there are more of them, and they form more quickly.

Artificial Ants

Myrmecology, the study of ants, might seem like a rather narrow and specialized scientific domain. But a growing number of researchers from outside the tight-knit myrmecology community have begun to take an interest in ants.

References to ants show up in unlikely places, from unlikely sources. In *Gödel, Escher, Bach,* Douglas Hofstadter (1979) describes a fictitious ant colony that he punningly calls Aunt Hillary. Hofstadter uses Aunt Hillary to explore differences between levels—in particular, differences between an ant colony as a whole and the individual ants that compose it. According to Hofstadter the workings of an ant colony can serve as a rough metaphor for the workings of the human brain: in each case, the behavior of the whole (colony or brain) is far more sophisticated (and of a very different character) than the behaviors of the component parts (ants or neurons).

In the nascent artificial life (ALife) community, dozens of researchers are creating simulations of artificial ants (for example, Collins and Jefferson 1991; Deneubourg and Goss 1989; Steels, 1990; Travers 1989). Indeed, ants have become the unofficial mascots of the ALife community. Artificial life posters are frequently illustrated with drawings of ants, and participants at ALife conferences adorn their name badges with plastic ants.

Interest in ants has even spread to the popular culture. Although not yet as popular as fish aquariums, ant farms are now becoming increasingly common in American homes. Uncle Milton Industries has sold more than 13 million of its Uncle Milton's Ant Farms, populated with 200 million *Pogonomyrmex californicus* ants (Miller 1991). Thousands of other households play with ants on their computer screens, using SimAnt software from Maxis. And *The Ants,* the definitive ant reference book by Harvard myrmecologists Bert Holldobler and E. O. Wilson (1990), has become a sort of cult classic, attracting attention far outside the myrmecology community.

Why the growing interest in ants? Many people, it seems, are intrigued with the collective nature of ant behavior. Each individual ant is quite simple. But an ant colony as a whole is capable of rather sophisticated behavior. Thus ant colonies have come to be viewed as a prototypical example of how complex-group behavior can arise from simple-individual behavior. As such, many people see the colony/ant relationship as an illuminating model (or, at least, an inspiring metaphor) for thinking about other group/individual relationships—such as the relationship between an organ and its cells, a cell and its macromolecules, a corporation and its employees, or a country and its citizens.

Compared with these other collective systems, ant colonies have the advantage of being more easily studied. As ant researchers Jean-Louis Deneubourg and Simon Goss (1989) note, "We can experiment on these [ant] societies in a way impossible in any other kind of collective decision-making organization. Unlike molecules or cells, [ant] workers are easily visible, and we can manipulate insect societies and place them in experimentally controllable situations with relative ease." Artificial ants, simulated on the computer screen, are even more easily manipulated and controlled (though at a risk of violating biological or physical realism). In particular, artificial ant simulations can help researchers probe the *mechanisms* that underlie collective behaviors. By experimenting with artificial ants, researchers can explore which individual-ant behaviors give rise to which colony-level behaviors.

Research on collective behavior in ant colonies has focused primarily on foraging activity (that is, how ants find and collect their food). Different species of ants search for food in different ways. In some species ants forage individually and adjust their strategies based on experience. But most species forage collectively, helping one another find and gather food. Typically, ants use *recruitment* strategies. When an ant finds some food, it recruits other ants to the food, who in turn recruit other ants to the food, and so on. The process slows down when there are fewer ants left to be recruited (or when there are other forces competing for the ants' attention).

Recruitment can take several different forms. In some species ants communicate directly with other ants. After an ant finds some food, it returns to the nest and leads one or more nestmates back to the food source. In other species ants communicate indirectly, through a chemical pheromone—a process known as *mass recruitment*. After an ant finds some food, it returns to the nest, dropping a chemical pheromone as it walks.

(Different types of ants find their way back to the nest in different ways: some by memory, some by smell, some by visual cues.) When other ants detect the pheromone trail, they follow it to the food source. Then they, too, return to the nest, reinforcing the pheromone trail. Before long, there is a strong trail between the food and nest, with hundreds of ants walking back and forth.

What happens when the ants finish exploiting the entire food source? Ants drop the chemical pheromone only when they are carrying food. So when the food source is fully depleted, the ants no longer drop pheromone. The pheromone trail becomes weaker and weaker through evaporation. As the trail becomes weaker, the ants become less likely to follow it. Instead they wander off in search of a new food source.

This mass recruitment process is implemented in the following StarLogo program. Each ant's actions are controlled by four demons. One demon tells the ant how to look for food (follow the pheromone if you sense it, wander randomly if you don't). A second demon tells the ant what to do when it finally bumps into the food (pick up a piece of food and turn around). A third demon tells the ant how to return to the nest (follow the scent of the nest, dropping pheromone as you go). And the fourth demon tells the ant what to do when it gets back to the nest (drop the food, and turn around to go get more). Meanwhile, the patches indicate where the nest and food are, and they cause the pheromone to evaporate and diffuse. (See the programming notes, following the program, for more details.)

Observer Procedures

```
to setup
clear-all
create-turtle 100
turtle-setup
patch-setup
end
```

Turtle Procedures

```
to turtle-setup
turtles-have [carrying-food? drop-size]
setxy 0 0
set-sniff-distance 3.0
set-carrying-food? false
activate-demon [look-for-food-demon
                find-food-demon
                return-to-nest-demon
                find-nest-demon]
end

to look-for-food-demon
if not carrying-food?
   [ifelse (ask patch-here [pheromone]) < 0.2
       [right random 40 left random 40]
       [set-heading uphill pheromone]
     forward 1]
end

to find-food-demon
if (not carrying-food?)
      and ask patch-here [food > 0]
   [set-carrying-food? true
    ask patch-here [set-food food - 1]
    set-drop-size 35
    right 180 forward 1]
end

to return-to-nest-demon
if carrying-food?
   [ask patch-here [add-pheromone-drop]
    set-drop-size drop-size - 0.6
    set-heading uphill nest-scent
    forward 1]
end

to find-nest-demon
if carrying-food? and ask patch-here [nest?]
    [set-carrying-food? false
     right 180 forward 1]
end
```

Patch Procedures

```
to patch-setup
patches-have [pheromone food nest? nest-scent]
set-pheromone 0
set-diffusion-rate 0.15
setup-nest
setup-food
activate-demon [diffuse-demon evaporate-demon
                update-colors-demon]
end

to setup-nest
ifelse (distance 0 0) < 5
   [set-nest? true
    set-nest-scent 1000]
   [set-nest? false
    set-nest-scent 0]
repeat 100 [diffuse nest-scent]
end

to setup-food
ifelse (distance 30 0) < 4
   [set-food 1 + random 3]
   [set-food 0]
end

to diffuse-demon
diffuse pheromone
end

to evaporate-demon
set-pheromone pheromone * 0.95
end

to update-colors-demon
ifelse nest?
   [set-patchcolor purple]
   [ifelse food > 0
      [set-patchcolor blue]
      [scale-patchcolor green pheromone 0 2]]
end

to add-pheromone-drop
set-pheromone pheromone
              + ask turtle-here [drop-size]
end
```

Programming Notes

To set up the nest-scent field, the program places a high level of nest-scent in the nest, then diffuses it one hundred times. The diffusion creates a radially symmetric "hill" of nest-scent, with the nest at the peak of the hill. By following the gradient of nest-scent, ants return directly to the nest.

The ants' sniff-distance is set to 3.0. So ants sniff three patches ahead when they are following the gradient. Setting sniff-distance greater than its default of 1.0 effectively increases each turtle's size in relation to the patches (or, conversely, increases the resolution of the patches in relation to the turtles).

Ants drop decreasing amounts of pheromone as they return to the nest (controlled by the variable drop-size). That way, the gradient of the trail is biased toward the food. So when another ant finds the trail, it is more likely to follow the trail toward the food, not the nest.

When the program is run, a hundred ants stream out of the nest, searching (randomly) for food (figure 3.4a–b). Once an ant finds some food, it brings a piece back to the nest, laying a green pheromone trail as it returns (figure 3.4c). At first, the green trail is thin and faint. But as other ants follow the pheromone to the food, and reinforce the trail on their way back to the nest, a thick, bright green trail develops (figure 3.4d). The pheromone trail is an example of a large-scale, orderly structure, created entirely through local interactions. It lasts only as long as the food source. Once the food source is fully exploited, the green trail gradually fades away (figure 3.4e), and the ants wander around aimlessly (figure 3.4f).

This program can be extended in several ways. For example, we could add two new food sources, each at a different distance from the nest. (The new food sources can be added with a few simple changes to the setup-food procedure.) The resulting colony-level behavior is quite striking (figure 3.5). The colony exploits the food sources as if controlled by a centralized plan. The ants initially exploit the source closest to the nest, then (after that source is fully depleted) they begin to exploit to the next closest source, and so on. It is as if the some leader in the nest had developed a plan for collecting the food systematically and sequentially.

But of course there is no leader. The high-level, sequential behavior arises entirely from low-level, parallel interactions.

What causes this sequential, planlike behavior? Here's one way to think about it. For each food source, there is a critical density of ants needed to form a solid, stable trail. If the number of ants is below critical density, a trail will diffuse and evaporate more quickly than it is reinforced by the ants. The critical density depends on the distance of the food source from the nest (along with other factors, like the evaporation and diffusion rates). The more distant a food source, the higher the critical density. Why? For more distant food sources, each ant takes a longer time to travel between food and nest, so it reinforces the trail less often. As a result, more ants are needed to counterbalance the forces of diffusion and evaporation.

So what happens when ants are released in an environment with three food sources? In some ways, it is helpful to think about the food sources

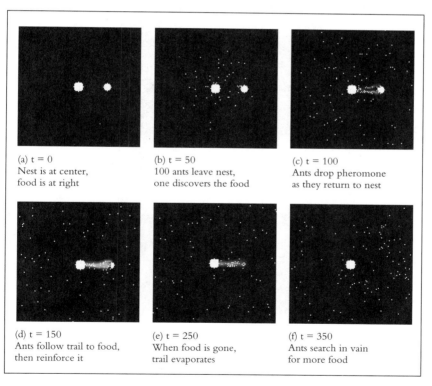

(a) t = 0
Nest is at center,
food is at right

(b) t = 50
100 ants leave nest,
one discovers the food

(c) t = 100
Ants drop pheromone
as they return to nest

(d) t = 150
Ants follow trail to food,
then reinforce it

(e) t = 250
When food is gone,
trail evaporates

(f) t = 350
Ants search in vain
for more food

Figure 3.4
Ant foraging with a single food source

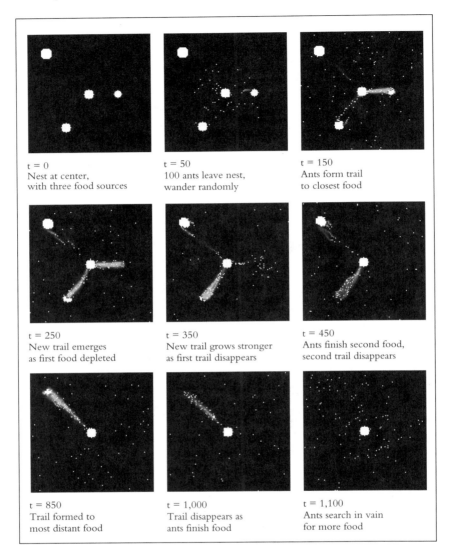

Figure 3.5
Ant foraging with three food sources (nest at center)

as competitors, each trying to attract a stable trail of ants. In this competition the food source closest to the nest has two advantages: it is the one most likely to be discovered in a random walk from the nest, and it has the lowest critical density (that is, it needs the smallest number of ants to form a stable pheromone trail). So as the ants march out of the nest and explore the world, the closest food source is most likely to win the competition. That is, the colony's first stable trail is most likely to go to the closest food source.

Once an ant joins a stable trail, it is unlikely to leave the trail, as long as the food remains. So ants on the stable trail are taken out of circulation. Assuming a fixed supply of ants (as in the StarLogo program), there are fewer free ants remaining to explore and form trails to the other food sources. Other food sources are considerably less likely to attract the critical density of ants needed to create a trail. The colony is likely to form only one strong trail at a time. But once the closest food source is fully depleted, the situation changes. The pheromone trail to that source dissipates, freeing the ants that had been gathering food along that trail. After that, there are again enough ants to form a new trail. By the same reasoning as before, the colony is most likely to form a trail to the closest of the remaining food sources.

And so the process goes. At any given time the colony is most likely to form a trail to the closest food source. And while the colony is exploiting that food source, it is unlikely to form trails to any of the other food sources. In this way the colony exploits the food sources one by one, in a seemingly planned fashion, moving outward from the closest food source to the most distant.

What happens if two food sources are equidistant from the nest? Would the colony form a weak trail to each of the food sources? Or would the ants form trails to neither, somewhat like the mythical "perfectly rational donkey" that starved when it couldn't decide between two equally distant piles of hay? In fact, a real ant colony (of sufficient size) would most likely choose between the two food sources, focusing most of its resources on one of the two sources (Deneubourg et al. 1986).

The StarLogo program exhibits the same behavior, closely mimicking the behavior of a real ant colony (Robson and Resnick 1991). The colony makes its choice through a positive feedback mechanism. Initially the ants make weak trails, of roughly equal strength, to each of the food sources. But any difference in the strengths of the two trails, caused by random factors, is quickly accentuated. Once one trail is slightly stronger, it is slightly more likely to attract free ants, and it is slightly less likely to lose the ants that are already on the trail. On the weaker trail, ants are slightly more likely to wander off the trail. So the stronger trail is likely to get stronger still, and the weaker trail, weaker. In this way, small differences can grow quickly, leading to a "symmetry breaking" in which one trail becomes dominant. Deneubourg et al. (1986) speculate that such symmetry breaking could have evolutionary advantages for ants, "as it allows the society to concentrate the exploitation on one source which is then better defended than if the foragers were divided around several sources."

Traffic Jams

At the time Ari and Fadhil started working with StarLogo, they were also taking a driver's education class. Each had turned 16 years old a short time before, and they were excited about getting their driver's licenses. Much of their conversation focused on cars. So when I gave Ari and Fadhil a collection of articles to read, it is not surprising that a *Scientific American* article titled "Vehicular Traffic Flow" (Herman and Gardels 1963) captured their attention.

Traffic flow is rich domain for studying collective behavior. Interactions among cars in a traffic flow can lead to surprising group phenomena. Consider a long road with no cross streets or intersections. What if we added some traffic lights along the road? The traffic lights would seem to serve no constructive purpose. It would be natural to assume that the traffic lights would reduce the overall traffic throughput (number of cars per unit time). But in some situations, additional traffic lights actually *improve* overall traffic throughput. The New York City Port Authority, for example, found that it could increase traffic throughput in the Holland Tunnel by 6 percent by deliberately stopping some cars before they entered the tunnel (Herman and Gardels 1963).

Or imagine what would happen if a street were temporarily closed in a congested downtown area. With one less street, it would be natural to assume that traffic conditions would worsen. But assumptions would be wrong again. The New York City Transportation Commissioner found that closing 42nd Street actually improved traffic flow in the area (Cohen and Kelly 1990).

Ari and Fadhil, based on their membership in the teenage-driver culture, were familiar with a variety of interesting traffic behaviors. They told me about a phenomenon called "snaking," which I had never heard of.

Mitchel: *What's snaking?*

Fadhil: *First one lane slows down. Then the other lane slows down, and the first one starts moving. Then the first one slows down again, and the other one starts moving. So it's like a snake.*

Mitchel: *Why does it happen?*

Fadhil: *A lot of people switch from the first lane to the other lane. Once they get to the other lane, the first lane starts going again. It keeps going back and forth.*

Mitchel: *Because people keep switching?*

Fadhil: *You know how when you're in a lane. Then you switch to the next lane since you see it's going faster. Then that one stops. And you say, "Gee, I have no luck. All the lanes I go to slow down."*

Mitchel: *If you were the only car to switch, it would work, right?*

Fadhil: *Yeah. But since you're not, it would be best to stay in your lane.*

Traditional studies of traffic flow rely on sophisticated analytic techniques (from fields like queuing theory). But many of the same traffic phenomena can be explored with simple StarLogo programs. To get started, Ari and Fadhil decided to create a one-lane highway. (Later, they experimented with multiple lanes, to explore the snaking phenomenon.) Ari suggested adding a police radar trap somewhere along the road, to catch cars going above the speed limit. But he also wanted each car to have its own radar detector, so cars would know to slow down when they approached the radar trap. (Ari noted that his mother's car had a sophisticated radar detector, with "three levels of warnings.")

After some discussion, Ari and Fadhil decided that each StarLogo turtle/car should follow three basic rules:

• If there is a car close ahead of you, slow down.

• If there aren't any cars close ahead of you, speed up (unless you are already moving at the speed limit).

• If you detect a radar trap, slow down.

These rules can be implemented with three StarLogo demons: one to make the car move, one to check if another car is close ahead (and adjust the car's speed accordingly), and one to check for the radar trap (and adjust the car's speed if necessary).

Observer Procedures

```
to setup
clear-all
create-turtle 15
turtle-setup
patch-setup
end
```

Turtle Procedures

```
to turtle-setup
turtles-have [step-size speed-limit
              danger-distance]
set-heading 90
sety 0
set-step-size 0.5 + (0.1 * random 16)
set-speed-limit 3
set-danger-distance 4
activate-demon [move-demon collision-demon
                radar-demon]
end

to move-demon
forward step-size
end

to collision-demon
ifelse car-ahead? danger-distance
   [set-step-size 0.5]
   [set-step-size step-size + 0.1]
if step-size > speed-limit
   [set-step-size speed-limit]
end

to radar-demon
if (ask patch-here [radar?])
    and (step-size > 0.8)
   [set-step-size step-size - 0.4]
end

to car-ahead? dist
ifelse dist <= 0
   [false]
   [ifelse ask patch-polar dist heading
             [turtle-total > 0]
       [true]
       [car-ahead? dist - 1]]
end
```

Patch Procedures

```
to patch-setup
patches-have radar?
set-patchcolor green
if patch-ypos = 0
   [set-patchcolor black]
setup-radar
end

to setup-radar
if (patch-ypos = 0) and (patch-xpos > 20)
                     and (patch-xpos < 25)
   [set-radar? true
    set-patchcolor blue]
   [set-radar? false]
end
```

Programming Notes

Note that cars decelerate abruptly (to a step-size of 0.5) when they see other cars ahead. Initially, Ari and Fadhil made cars decelerate more gradually, but cars sometimes "jumped over" other cars. So they changed the program to make the cars decelerate more abruptly.

Ari and Fadhil expected that a traffic jam would form behind the radar trap, and indeed it did (figure 3.6). After a few dozen iterations of the StarLogo program, a line of cars started to form to the left of the blue radar trap. The cars moved slowly through the trap, then sped away as soon as they passed it. Ari explained, "First one car slows down for the radar trap, then the one behind it slows down, then the one behind that one, and then you've got a traffic jam." The only unexpected effect was the rapid acceleration of the cars as they moved beyond the radar trap. The radar trap, in effect, organized the cars for maximum acceleration. As the cars slowed down for the radar trap, they formed a queue with roughly equal distances between each car. So when the cars moved beyond the radar trap, they did not interfere with one another. The cars were "released" by the radar trap one by one, and they accelerated smoothly until they reached the speed limit.

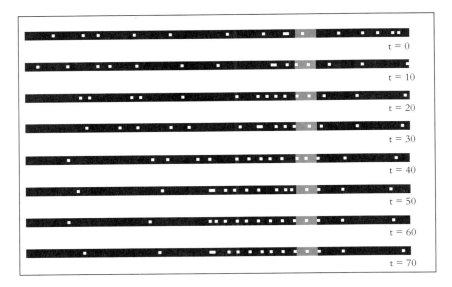

Figure 3.6
Traffic jam caused by radar trap (cars move left to right)

Figure 3.7
Traffic jam without radar trap (cars move left to right, but jam moves right to left)

Ari and Fadhil noted that an accident on the side of the road would have the same effect as the radar trap, due to the infamous "rubbernecking" effect. In fact, they argued that even a tiny disruption could cause a jam. Fadhil explained, "When a car on the highway even touches the brakes, the brake lights go on. Even if it doesn't slow down, everyone else slows down. If the first person just touches the brake, the brake lights go on, and the person behind him doesn't want to hit him so he slows down a little bit more, and the person behind him a little bit more, and the person behind him more, and you end up having a traffic jam. And the first guy didn't even slow down at all."

I asked Ari and Fadhil what would happen if only *some* of the cars had radar detectors. Ari predicted that only some of the cars would slow down for the radar trap. Fadhil had a different idea: "The ones that have radar detectors will slow down, which will cause the other ones to slow down." Ari found that argument compelling and quickly changed his mind. And, indeed, Fadhil was right. We modified the StarLogo program so that only 25 percent of the cars had radar detectors (that is, only 25 percent of the cars had the `radar-demon` activated). The result: the traffic flow looked exactly the same as when all of the cars had radar detectors.

What if *none* of the cars had radar detectors—or, equivalently, if the radar trap were removed entirely? With no radar trap, the cars would be controlled by just two simple rules: if you see another car close ahead, slow down; if not, speed up. The rules couldn't be much simpler. At first, Fadhil predicted that the traffic flow would become uniform: cars would be evenly spaced, traveling at a constant speed. After all, without the radar trap, what could cause a jam? But he quickly changed his mind and predicted that a traffic jam would form.

In fact, when we ran the program, a traffic jam formed (figure 3.7). Along parts of the road, the cars were tightly packed and moving slowly. Elsewhere, they were spread out and moving at the speed limit. Watching one of the tightly packed jams, Fadhil was reminded of the toy with five pendulum balls in a row. When a pendulum ball strikes one end of the pack, a ball shoots out from the other end. The traffic jam looked somewhat similar: "Whenever one car leaves [the jam], another one comes in, so it [the jam] keeps the same amount. . . . Then another will get caught up and another will leave. Caught, leave, caught, leave."

Fadhil thought that the jams were caused by differences in the initial speeds of the cars. So we changed the StarLogo program, starting all of the cars at the exact same speed. But the jams still formed. Fadhil quickly understood. At the beginning of the program, the cars were placed at random positions on the road. Random positioning led to uneven spacing between the cars, and uneven spacing could also provide the seed for a traffic jam to form. Fadhil explained, "Some of the cars start closer to other cars. Like, four spaces between two of them, and two spaces between others. A car that's only two spaces behind another car slows down, then the one behind it slows down."

Next, we changed the program so that the cars were evenly spaced. Sure enough, no traffic jams formed. All of the cars uniformly accelerated up to the speed limit. But Ari and Fadhil recognized that such a situation would be difficult to set up in the real world. The distances between the cars had to be just right, and the cars had to start at exactly the same time—like a platoon of soldiers starting to march in unison.

We reintroduced some randomness in the initial conditions of the StarLogo program, and the traffic jams returned. Watching the traffic jams more closely, Ari and Fadhil noticed that the jams did not stay in one place but tended to move with time. In fact, the traffic jams tended to move *backward,* even though all of the cars within them were moving forward. Fadhil described it: "The jam itself moves backward. If you keep your eye on one car, it leaves the traffic jam, but the jam itself, I mean where you see the cars piling up, moves backward."

The backward movement of the traffic jam highlights an important idea: collective structures (like traffic jams) often behave very differently from the elements that compose them. This idea is true not only for traffic jams but for a much wider range of phenomena, including waves. Ideas about waves are very difficult for students to grasp. One reason is that waves are often presented in unmotivated contexts (for example, perturbations moving along a string) and through difficult mathematical formalisms (such as differential equations).

StarLogo seems to provide a more accessible introduction to wavelike phenomena. Like differential equations, StarLogo can be used as a formal system for expressing ideas about wave behavior. But the StarLogo representation is different in several important ways. For one thing, StarLogo programs seem easier to understand and manipulate. In addition, StarLogo programs are executable, so that students can watch their programs run and revise their ideas based on what they see. Perhaps most important, StarLogo offers students a chance to explore wave phenomena in personally meaningful contexts. The fact that Ari and Fadhil developed strong intuitions about traffic flow while working on their StarLogo project was due, in no small part, to their deep interests in and experiences with cars.

Termites

Philip Morrison, the MIT physicist and science educator, once told me a story about his childhood. When Morrison was in elementary school, one of his teachers described the invention of the arch as one of the central, defining milestones of human civilization. Arches took on a special meaning for the young Morrison. He felt a certain type of pride whenever he saw an arch. Many years later, when Morrison learned that lowly termites also build arches, he was quite surprised (and amused). He gained a new skepticism about everything that he was taught in school, and a new respect for the capabilities of termites. Ever since, Morrison has wondered about the limits of what termites might be able to do. If they can build arches, why not more complex structures? Given enough time, Morrison wondered, might termites build a radio telescope?

Probably not. But termites *are* among the master architects of the animal world. On the plains of Africa, termites construct giant moundlike nests rising more than ten feet tall, thousands of times taller than the termites themselves. Inside the mounds are intricate networks of tunnels and chambers. Certain species of termites even use architectural tricks to regulate the temperature inside their nests, in effect turning their nests into elaborate air-conditioning systems. As E. O. Wilson (1971) notes, "The entire history of the termites . . . can be viewed as a slow escape by means of architectural innovation from a dependence on rotting wood for shelter" (p. 315).

Each termite colony has a queen. But, as in ant colonies, the termite queen does not "tell" the termite workers what to do. (In fact, it seems fair to wonder if the designation "queen" is a reflection of human biases. "Queen" seems to imply "leader." But the queen is more of a mother to the colony than a leader.) On the termite construction site there is no construction foreman, no one in charge of the master plan. Rather, each termite carries out a relatively simple task. Termites are practically blind, so they must interact with each other (and with the world around them) primarily through their senses of touch and smell. But from local interactions among thousands of termites, impressive structures emerge.

The global-from-local nature of termite constructions makes them well suited for StarLogo explorations. Of course, simulating the construction of an entire termite nest would be a monumental project (involving many details unrelated to my interests). So I chose a far more modest goal: program some artificial termites to collect wood chips and put them into piles. (Real termites don't actually carry wood chips from place to place. Rather, they eat pieces of wood, then build structures with "fecal cement" that they produce from the digested wood.)

The challenge is to figure out a decentralized strategy for adding some order to a disordered collection of wood chips. At the start of the program, the wood chips are scattered randomly throughout the termites' world. But as the program runs, the termites should organize the wood chips into a few, orderly piles.

Someone suggested a very simple strategy to me. He suggested that each individual termite should obey the following rules:

• If you are not carrying anything and you bump into a wood chip, pick it up.

• If you are carrying a wood chip and you bump into another wood chip, put down the wood chip you're carrying.

At first, I was skeptical. While I am well aware of the power of simple rules, these rules seemed a bit too simple. There was no mechanism for preventing termites from taking wood chips away from existing piles. So while termites are putting new wood chips on a pile, other termites might be taking wood chips away from it. It seemed like a good prescription for getting nowhere.

But, putting my skepticism aside (or perhaps as a test of my skepticism), I decided to implement this simple strategy in StarLogo, to see what it would do. I asked one of the high-school students, Callie, if she wanted to work on the project with me, and she agreed.

When I described the simple rules to Callie, she was even more skeptical than I was. She explained, "There might not be any mounds, because every time a termite puts it [a wood chip] down, somebody might pick it up." But we pushed ahead with the StarLogo program anyway. We wrote four demons for the turtles/termites: one to make them move forward, another to make them "wiggle" a bit as they move, a third to make them pick up wood chips, and a fourth to make them put down wood chips.

Observer Procedures

```
to setup
clear-all
create-turtle 1000
turtle-setup
patch-setup
activate-demon observer-demon
end

to observer-demon
every 50 [print patch-subtotal [wood-chips > 0]]
end
```

Turtle Procedures

```
to turtle-setup
activate-demon [move-demon wiggle-demon
                look-for-chip-demon
                look-for-pile-demon]
end

to move-demon
forward 1
end

to wiggle-demon
right random 50
left random 50
end

to look-for-chip-demon
if (not carrying-wood-chip?)
    and any-wood-chips-here?
   [pick-up-chip
    right 180 forward 1]
end

to look-for-pile-demon
if carrying-wood-chip? and any-wood-chips-here?
   [put-down-chip
    right 180 forward 1]
end

to pick-up-chip
ask patch-here [set-wood-chips wood-chips - 1]
set-color blue
end

to put-down-chip
ask patch-here [set-wood-chips wood-chips + 1]
set-color red
end

to any-wood-chips-here?
ask patch-here [wood-chips > 0]
end

to carrying-wood-chip?
color = blue
end
```

Patch Procedures

```
to patch-setup
patches-have wood-chips
ifelse 0 = (random 8)
    [set-wood-chips 1]
    [set-wood-chips 0]
activate-demon patch-color-demon
end

to patch-color-demon
scale-patchcolor yellow wood-chips 0 10
end
```

Programming Notes

In the procedures look-for-chip-demon and look-for-pile-demon the termites turn around and take a step (right 180 forward 1) after they pick up or put down wood chips. They do so in order to minimize the chances of putting down a wood chip right after picking it up, or picking up a wood chip right after putting one down.

The observer-demon procedure prints the number of piles every 50 time steps. It is an example of how demons can be used to monitor a process.

The patch-color-demon procedure colors each patch a different shade of yellow, based on the number of wood chips on the patch. The more wood chips, the brighter the intensity of yellow. Patches without any wood chips are black. Patches with ten or more wood chips are high-intensity yellow.

In the initial version of this program, termites put down their wood chips *next to* other wood chips, rather than *on top of* other wood chips. That program was visually more interesting, since piles spread out horizontally on the screen, rather than piling up on a single patch. But the behavior of the termites was somewhat more difficult to analyze. In particular, piles were not clearly defined. At times, termites picked up wood chips from the "middle" of a pile in such a way that what remained could be seen as either one large pile or two smaller ones.

We tried the program, and (much to our surprise) it worked quite well. At first, the termites gathered the wood chips into hundreds of small piles. But gradually the number of piles declined, and the number of wood chips in each pile increased (see figure 3.8). After 2,000 iterations, there were about 100 piles, with an average of 15 wood chips in each pile. After 10,000 iterations, there were fewer than 50 piles left, with an average of 30 wood chips in each pile. After 20,000 iterations, only 34 piles remained, with an average of 44 wood chips in each pile. The process was rather slow. And it was frustrating to watch, as termites often carried wood chips away from well-established piles. But, all in all, the program worked quite well.

Why did it work? As I watched the program, it suddenly seemed obvious. Imagine what happens when the termites (by chance) remove all of the wood chips from a particular pile. Because all of the wood chips are gone from that spot, termites will never again drop wood chips there. So the pile has no way of restarting. I explained this reasoning to Callie, and she understood immediately: "You can't really start new clusters, can you? If you see one, you put it down. But if you don't see anything, you can't put yours down!"

As long as a pile exists, its size is a two-way street: it can either grow or shrink. But the *existence* of a pile is a one-way street: once it is gone, it is gone forever. Thus a pile is somewhat analogous to a species of creatures in the real world. As long as the species exists, the number of individuals in the species can go up or down. But once all of the individuals

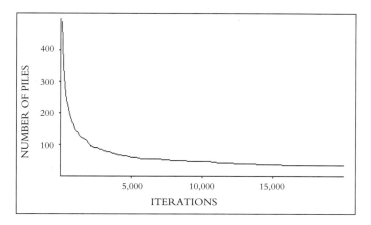

Figure 3.8
The number of wood-chip piles decreases monotonically

are gone, the species is extinct, gone forever. In these cases, zero is a "trapped state": once the number of creatures in a species (or the number of wood chips in a pile) goes to zero, it can never rebound.

Of course, the analogy between species and piles breaks down in some ways. New species are sometimes created, as offshoots of existing species. But in the termite program, there is no way to create a new pile. The program starts with roughly 2,000 wood chips. These wood chips can be viewed as 2,000 "piles," each with a single wood chip. As the program runs, some piles disappear, and no new piles are created. So the total number of piles keeps shrinking and shrinking.

Callie and I wanted to make the number of piles shrink even more quickly. So we added more termites. Instead of 1,000 termites, we used 4,000 termites. And indeed, the number of piles shrank much more quickly. With 1,000 termites, there had been 102 piles after 2,000 iterations, and 47 piles after 10,000 iterations. With 4,000 termites, there were only 10 piles after 2,000 iterations, and just 3 piles after 10,000. If we waited long enough, the wood chips would inevitably end up in a single pile.

But there is something a bit deceptive in these numbers. When the 4,000 termites gathered the wood chips into three piles, each pile had (on average) about 80 wood chips. That's about 240 wood chips. But we started with 2,000 wood chips. Where did the rest of them go? The termites were carrying them. By the symmetry of the situation, termites were equally likely to be carrying wood chips or not carrying wood chips. So at any time, roughly half of the termites were carrying wood chips. With 1,000 termites, about 500 wood chips were being carried, and about 1,500 were in piles. With 4,000 termites, almost all of the wood chips are being carried. Very few are actually in piles. In fact, when we ran the program with 8,000 termites, the termites picked up *all* of the wood chips. Then, the termites had nowhere to put the wood chips down. The number of piles dropped to zero and stayed there—another example of a trapped state.

So simply increasing the number of termites was not a very satisfactory solution. How else could we make the number of piles decrease more quickly? We decided to "protect" tall piles, so that termites would not destroy piles that had been well established. We changed the look-for-chip-demon procedure so that termites would pick up wood chips only from "short" piles—piles with nine or fewer wood chips. Piles with ten or more wood chips were protected; they could only grow, never shrink.

We ran the new program with 1,000 termites. As long as all wood-chip piles were short, the program ran exactly as before. But as soon as some piles accumulated more than ten wood chips, the dynamics of the activity changed. A symmetry was broken. Termites could put down wood chips on any pile, but they could pick up wood chips from only certain piles. So short piles were at a competitive disadvantage.

As a result, the number of piles declined more quickly than before. In the original program, without any protection for tall piles, the termites produced 157 piles after 1,000 iterations, each with an average of about 10 wood chips. With tall piles protected, there were only 85 piles after 1,000 iterations. As an added bonus, almost all of the wood chips were actually in piles; very few were being carried around by termites. After 1,000 iterations, nearly 1,900 of the 2,000 wood chips were in piles. The piles were (on average) more than 22 wood chips tall—more than twice as tall as in the "no protection" case.

But this rapid convergence to fewer and taller piles comes at a price. After 2,000 iterations, the number of piles had shrunk to 82. But all 82 piles had at least 10 wood chips. So all of the piles were protected. The termites could not find any more wood chips to pick up. The number of piles would stay frozen at 82 forever.

In the original program, with no protection for tall piles, the number of piles declined more gradually, but it kept declining. After 2,000 iterations, there were still more than 100 piles. But after 5,000 iterations, there were only 60 piles. After 10,000 iterations, only 47 piles. And the number of piles keeps declining. After 40,000 iterations, there were only 25 piles, each with an average of 60 wood chips.

The no-protection strategy seems inefficient, since it allows termites to take wood chips away from well-established piles. But in the long run, this strategy is more effective for gathering the wood chips into fewer piles. Moreover, Callie and I both preferred the "feel" of the no-protection program. Piles are never frozen in place. Piles continually shrink and grow, as if in a competition for survival. As Callie noted, "It feels more alive."

Turtles and Frogs

Once upon a time, in a land far away, there was a pond inhabited by turtles and frogs. It was a very happy pond: the turtles liked the frogs, and the frogs liked the turtles. They all got along swimmingly. For many years the turtles and frogs divided up the pond almost as if it were a checkerboard. On one lily pad lived a frog, on the next lily pad a turtle, on the next another frog, then another turtle, and so on. There was a nice symmetry to it. Each turtle had eight neighbors (including the corners): four turtles and four frogs. Similarly, each frog had four frog neighbors and four turtle neighbors.

Then, one dark night, a terrible storm hit the pond. Lightning cracked through the sky, and heavy rains beat down on the pond. A cold wind whipped along the surface of the pond, flipping lily pads through the air. Turtles and frogs were tossed all over the pond. Several of the creatures were killed as they crashed against the rocks.

By the next morning the rains had stopped, and the sun peeked through the clouds. The turtles and frogs surveyed the damage. The lily pads were scattered all over the pond, but luckily they were largely intact. The turtles and frogs spent some time rearranging the lily pads into a neat array, as they had been before. Then the creatures set out to find new places to live, each one looking for a lily pad that it could call home.

How did they choose among all of the lily pads? The lily pads were almost identical, so that wasn't a factor. And the creatures were quite tolerant: frogs didn't mind living next to turtles, and turtles didn't mind living next to frogs. But each turtle did want to make sure that there were at least some other turtles nearby. And similarly, each frog wanted to make sure that there were at least some other frogs nearby.

They had been happy with the previous arrangement, but they weren't quite sure how to re-create it. There was no one in charge to tell them where to go. So the turtles started crawling around (and the frogs started hopping around), hoping to find lily pads where they would be happy. Each turtle hoped to find a lily pad where at least 30 percent of its neighbors were turtles. If fewer than 30 percent of its neighbors were turtles, it would look for an empty lily pad nearby and move there, hoping to find more turtle neighbors. If 30 percent or more of the neighbors were turtles, it would settle in. But, of course, if the neighborhood changed, bringing the turtle percentage below 30 percent, it would have to start moving again. (Each frog followed an analogous strategy, hoping to find a lily pad where at least 30 percent of its neighbors were frogs.)

After a while, all of the creatures found lily pads where they were happy. Each creature had at least 30 percent of "its own kind" as neighbors. But the pond seemed quite different from before. Many of the turtles didn't have *any* frogs as neighbors, and many of the frogs didn't have *any* turtles as neighbors.

Franny Frog was one of the frogs without any turtle neighbors. She was very concerned and confused. She asked her friend Granny Goose if the pond, when viewed from above, looked much different from the way it looked before the storm. "It sure does," said Granny Goose. "I used to love looking down at the pond as I flew above. There was such a uniform mix of frogs and turtles. Now the pond seems so segregated. The frogs are living in some clusters, and the turtles are living in other clusters. In fact, I did a little calculation as I flew over the pond. I counted all of the neighbors for all of the turtles, and I found that more than 70 percent of the turtles' neighbors are other turtles. And it was the same for you frogs. More than 70 percent of your neighbors are other frogs. I figured that you frogs must have had a terrible fight with the turtles after the storm."

"I don't get it," sighed Franny Frog. "We didn't have a fight. We still like each other. We just wanted to have a few neighbors like ourselves. We didn't want to be split apart. What could have happened?"

If only Franny Frog had had access to StarLogo, she could have gained a better understanding of what happened. Below is a StarLogo program that re-creates the story of the turtles and the frogs.

Observer Procedures

```
to setup
clear-all
set-scale 128 64
create-breed frog
create-turtle 3000
create-frog 3000
creature-setup
end

to cluster-ratios-demon
every 20
   [type [Turtle cluster ratio:]
    print (turtle-sum [turtle-neighbors])
            / (turtle-sum [total-neighbors])
    type [Frog cluster ratio:]
    print (frog-sum [frog-neighbors])
            / (frog-sum [total-neighbors])]
end
```

Turtle Procedures

```
to creature-setup
creatures-have minimum-percentage
creatures-have test-distance
set-minimum-percentage 0.3
if turtle? [activate-demon turtle-move-demon]
if frog? [activate-demon frog-move-demon]
activate-demon avoid-doubling-demon
end

to turtle-move-demon
if turtle-neighbors
      < (minimum-percentage * total-neighbors)
   [find-free-patch]
end

to frog-move-demon
if frog-neighbors
      < (minimum-percentage * total-neighbors)
   [find-free-patch]
end

to avoid-doubling-demon
if ask patch-here [creature-total > 1]
   [find-free-patch]
end

to find-free-patch
set-heading 45 * random 8
set-test-distance 1 + random 5
if ask patch-polar test-distance heading
      [creature-total = 0]
   [forward-grid test-distance]
end

to turtle-neighbors
ask patch-here [neighbor-sum [turtle-total]]
end

to frog-neighbors
ask patch-here [neighbor-sum [frog-total]]
end

to total-neighbors
ask patch-here [neighbor-sum [creature-total]]
end
```

Programming Notes

create-breed frog creates a new "breed" of creature (named "frog") and automatically generates new primitives for that breed (create-frog, frog?, frog-sum, etc.) See appendix B for more details on StarLogo breeds.

By default, the size of the StarLogo world is 128 × 128 (16,384 patches). But this program sets the size of the "pond" to 64 × 128 (8,192 patches). That size forces the 6,000 creatures to keep in close proximity to one another but gives them enough room to move if needed.

A creature moves for one of two reasons: it is unhappy with its mix of neighbors (detected by the frog-move-demon or the turtle-move-demon), or it is sharing its patch with another creature (detected by the avoid-doubling-demon). In either case it calls the find-free-patch procedure. In this procedure the creature randomly picks a nearby patch. If the patch is empty, the creature moves there.

Let's consider one run of this program. The setup procedure scatters turtles and frogs randomly around the pond. There are 8,192 patches and 6,000 creatures (3,000 turtles and 3,000 frogs), so about one-quarter of the patches (2,192 out of 8,192) are empty. The turtles and frogs are quite well mixed throughout the pond. What I call the "turtle cluster ratio" is around 0.5. That means, of all the neighbors of all the turtles (throughout the whole pond), 50 percent are other turtles (and, of course, the other 50 percent are frogs). The "frog cluster ratio" is also around 0.5. So the frogs, on average, are also surrounded by half frogs and half turtles. (A frog cluster ratio of 1.0 would mean that frogs have only frogs as neighbors. A frog cluster ratio of 0.0 would mean that frogs have only turtles as neighbors.)

But individual turtles and frogs don't care about overall statistics. Each creature cares only about its own neighbors, and it wants 30 percent (or more) of those neighbors to be of its own kind. For example, if a turtle has four or five or six neighbors, it wants to be sure that at least two of them are turtles. If a turtle has seven or eight neighbors, it wants to be sure that at least three of them are turtles. In the initial setup, almost 18 percent of the creatures (524 turtles and 533 frogs) are unhappy with their mix of neighbors. These creatures start looking for new places to live. When an unhappy creature finds an empty patch nearby, it jumps there and hopes that the new patch has a better mix of neighbors.

The number of unhappy creatures does not necessarily decrease monotonically. When a creature jumps to a new patch, it changes the neighborhood mix for both its old neighbors and its new neighbors. Even if the just-moved creature is happy with its new neighbors, some of those neighbors might no longer be happy, and they will begin looking for new places to live. Although only 18 percent of the creatures were unhappy with their initial mix of neighbors, more than 34 percent of the creatures (1,016 turtles and 1,031 frogs) move sometime during the run of the program.

Nevertheless, the number of unhappy creatures tends to decrease with time. After 10 iterations of the program, only 6 percent of the creatures (177 turtles and 182 frogs) are still unhappy. After 20 iterations, only 3 percent (82 turtles and 81 frogs) are unhappy. After 40 iterations, only 1 percent (30 turtles and 34 frogs) are unhappy. Finally, after 113 iterations, all of the creatures are happy. No one moves any more.

But the growing percentage of happy creatures is achieved through increased segregation. Visually, the effect is clear. The random mix of turtles and frogs is gradually replaced by clusters of turtles and frogs. After 10 iterations, the turtle cluster ratio (which started at 0.5) has risen to 0.62—indicating that, on average, 62 percent of the turtles' neighbors are other turtles. The frog cluster ratio is similar. After 20 iterations, the cluster ratios (for both turtles and frogs) have moved to 0.67. By the time the system stabilizes, after 113 iterations, the cluster ratios have risen above 0.71. So while each turtle, individually, would be content with just 30 percent turtle neighbors, the turtles end up, on average, with 71 percent turtle neighbors.

This turtle/frog scenario was inspired by the writings of Harvard economist Thomas Schelling. In an article titled "On the Ecology of Micromotives," Schelling (1971) notes that the "micromotives" of individuals can lead to "macro" patterns that are not necessarily desired by any of the individuals. At a cocktail party, for instance, men and women might end up in single-gender conversation clusters, even if everyone would prefer mixed-gender clusters. And a residential neighborhood might become more segregated ethnically or racially than any individual would find desirable. As Schelling puts it, "A moderate urge to avoid small-minority status may cause a nearly integrated pattern to unravel and highly segregated neighborhoods to form." (Of course, this argu-

ment does not preclude the possibility that some individuals might actually *prefer* extreme segregation. Schelling's point is that extreme segregation can occur even in the absence of such extreme individuals.)

What's behind this segregation effect? It is best to start with a simple
example. Imagine a cocktail party with two clusters of people, each with
about ten people. Every once in a while, people shift from one cluster to
the other. Each cluster starts with a roughly equal number of men and
women. But as people move back and forth among the clusters, one of
the clusters ends up (by chance) with two men and eight women. One
of the men, feeling uncomfortable in such a small-minority status, drifts
off to the other cluster. By leaving, the man aggravates the situation that
caused him to leave in the first place. The cluster is now even more
dominated by women. The lone remaining man feels uncomfortable and
decides to leave, making the cluster even more dominated by women
and even more intimidating to men who might think of joining. And
the problem keeps getting worse. When the two men both join the
other cluster, they tip the balance of *that* cluster so that it becomes overly
dominated by men. The women in that cluster, feeling uncomfortable in
their new small-minority status, drift away, joining the women-dominat-
ed cluster that the two men just left.

In this case each person might prefer a mixed-gender cluster. But
once the ratio becomes lopsided in one of the clusters, there is no way
back. People in the minority begin to leave, unleashing a positive feedback loop that induces even more of the minority members to leave.
The result is two single-gender clusters.

In the turtle/frog program the situation is similar, but repeated many
times over. The initial distribution of turtles and frogs seems relatively
uniform. But there are some tiny regions dominated by one type of
creature or the other. Imagine a region dominated by frogs. When one
of the turtles in that region jumps to a new location, the ratio of frogs to
turtles becomes even more lopsided, inducing even more turtles to leave.
And as a turtle moves to a new location, it increases the turtle-to-frog
ratio in that region, perhaps inducing nearby frogs to look for new
homes. The ripple effect continue across the pond, until the frogs and
turtles are segregated into clusters.

If the individual creatures have an even greater desire to be near "their
own kind" (in terms of the StarLogo program, a higher value for the
variable `minimum-percentage`), the collective segregation is even
more pronounced. If each creature demands that at least 40 percent of its
neighbors be of the same kind (that is, if `minimum-percentage` is set

to .4), the creatures end up, on average, with 79 percent same-kind neighbors. If individuals demand at least 50 percent, the group average is 85 percent.

If individuals demand at least 60 percent same-kind neighbors, the group average is 95 percent. The turtles are almost entirely surrounded by other turtles, and frogs by other frogs. The solution looks quite different from previous solutions. The turtle clusters and frog clusters no longer abut one another; they are separated from one another by "moats" of empty space. It is almost as if the creatures, so intent on being near their own kind, decided to use the empty space to construct moats isolating themselves from other kinds of creatures.

If the individuals demand at least 70 percent, the group seems unable to settle down to a solution. After several thousand iterations of the program, the creatures seem no closer to a solution. The number of unhappy creatures remains high. Creatures keep jumping to new locations and disrupting other creatures.

Of course, all of these results are probabilistic. If each creature demands at least 50 percent same-kind neighbors, it is possible that the creatures could arrange themselves in a checkerboard. In that case all of the individual percentages (and hence the group percentage) would be precisely 50 percent. But the chances of the creatures arranging themselves into a checkerboard are infinitesimally small—like the chances of half a bathtub freezing while the other half boils. When each individual creature seeks at least 50 percent same-kind neighbors, the creatures almost inevitably settle into a pattern in which the group average of same-kind neighbors is near 85 percent.

Turtle Ecology

The great baseball manager Casey Stengel once said, "If you don't know where you're going, you might end up somewhere else." My experiences with computer-based explorations have taught me a corollary: "Even if you think you know where you're going, you'll probably end up somewhere else."

That happened to Benjamin, a student at Woburn High School, when he set out to create a StarLogo program that would simulate evolution by natural selection. In the collection of papers that I had given to the high-school students was a *Scientific American* article (Dewdney 1989) about a computer program called Simulated Evolution (Palmiter 1989). Benjamin, who had just finished his junior year in high school, read the article and decided that he wanted to create a StarLogo program similar to the commercial program described in the article. His goal was to create a set of computer creatures that would interact and evolve.

At the core of Benjamin's simulation were turtles and food. His basic idea was simple: turtles that eat a lot of food reproduce, and turtles that don't eat enough food die. Eventually, he planned to add "genes" to his turtles. Different genes could provide turtles with different levels of "fitness" (perhaps different capabilities for finding food). But Benjamin never got around to the genes. Rather, on the road to evolution, Benjamin got sidetracked into an interesting exploration of simple ecological systems.

Benjamin began by making food grow randomly throughout the StarLogo world. (During each time step, each StarLogo patch had a random chance of growing some food.) Then he created some turtles. The turtles had very meager sensory capabilities. They could not "see" or "smell" food at a distance. They could sense food only when they bumped directly into it. So the turtles followed a very simple strategy: wander around randomly, eating whatever food you bump into.

Benjamin gave each turtle an "energy" variable. Every time a turtle took a step, its energy decreased a bit. Every time it ate some food, its energy increased. Then Benjamin added one more rule: if a turtle's energy dipped to zero, the turtle died.

These ideas are captured in the following StarLogo program. The turtles' behavior is controlled by three demons. One makes the turtles move, another makes the turtles eat (if there is food), and a third makes the turtles die (if their energy falls to zero).

Observer Procedures

```
to setup number-of-turtles
clear-all
setup-food
create-turtle number-of-turtles
turtle-setup
end

to monitor-demon
every 20 [print turtle-total]
end
```

Turtle Procedures

```
to turtle-setup
turtles-have energy
set-energy 20
activate-demon [walk-demon eat-demon die-demon]
end

to walk-demon
right random 50
left random 50
forward 1
set-energy energy - 0.1
end

to eat-demon
if ask patch-here [food > 0]
   [ask patch-here [set-food 0]
    set-energy energy + 1]
end

to die-demon
if energy <= 0 [die]
end
```

Patch Procedures

```
to setup-food
ifelse (random 20) = 1
   [set-food 1 set-patchcolor green]
   [set-food 0 set-patchcolor black]
activate-demon food-demon
end

to food-demon
if (random 1000) = 1
   [set-food 1]
ifelse food = 0
   [set-patchcolor black]
   [set-patchcolor green]
end
```

Programming Notes

The monitor-demon keeps track of the total number of turtles alive. It prints out the total turtle population every 20 time steps.

With this program, the turtles do not reproduce. Life is a one-way street: turtles die, but no new turtles are born. Still, even with this simple-minded program, Benjamin found some surprising and interesting behaviors.

Benjamin ran the program with 300 turtles. But the environment could not support that many turtles. There wasn't enough food. So some turtles began to die. The turtle population fell rapidly at first, then it leveled out at about 150 turtles. The system seemed to reach a steady state with 150 turtles: the number of turtles and the density of food both remained roughly constant.

Then Benjamin tried the same program with 1,000 turtles. If there wasn't enough food for 300 turtles, there certainly wouldn't be enough for 1,000 turtles. So Benjamin wasn't surprised when the turtle population began to fall. But he *was* surprised with how *far* the population fell. After a while, only 28 turtles remained. Benjamin was puzzled: "We started with more, why should we end up with less?" After some discussion, he realized what had happened. With so many turtles, the food shortage was even more critical than before. The result: mass starvation.

I noted that many of the turtles had died right around the same time. I guessed that these turtles had eaten almost no food. They died when their initial energy supply ran out. I suggested a small change. Rather than each turtle starting with 20 units of energy, what if each turtle started with a random amount of energy less than 20 units? (That required only a small change in the `turtle-setup` procedure: changing 20 to `random 20`.) Although the overall turtle population would start with less energy, might more turtles survive in the long run?

Benjamin immediately grasped the idea. He explained, "The ones that die fast, the ones with less (initial) energy, leave more food. They won't waste the food by eating it and just dying." Benjamin predicted that more turtles would survive in the long run. And, sure enough, 97 turtles survived (compared with only 28 before). Benjamin understood what had happened, but he still found the behavior a bit strange: "The turtles have less [initial energy as a group], and less usually isn't more."

Next Benjamin decided to add reproduction to his model. His plan: whenever a turtle's energy increases above a certain threshold, the turtle should "clone" itself, and split its energy with its new twin. That can be accomplished by adding another demon procedure to the program.

Turtle Procedures

```
to clone-demon
if energy > 15
    [set-energy 0.5 * energy
     clone []]
end
```

Programming Notes

Before a turtle clones itself, it cuts its energy in half. That way, after cloning, both it and its clone will have half of the original energy.

The `turtle-setup` procedure should also be modified, so that it activates the `clone-demon` along with the other demons.

Benjamin assumed that the rule for cloning would somehow balance the rule for dying, leading to some sort of equilibrium. He explained, "Hopefully, it will balance itself out somehow. I mean it will. It will have to. But I don't know what number it will balance out at." After a little more thought, Benjamin suggested that the food supply might fall at first, but then it would rise back and become steady: "The food will go down, a lot of them will die, the food will go up, and it will balance out."

Benjamin started the program running. As he expected, the food supply went down and then up. But it didn't balance out: it went down and up again, and again, and again. Meanwhile, the turtle population also oscillated, but out of phase with the food. These dueling oscillations are characteristic of situations with predators (in this case, turtles) and prey (in this case, food).

Usually, scientific (and educational) explorations of predator-prey models are based on sets of differential equations, known as the Lotka-Volterra equations (Lotka 1925; Volterra 1926). For example, the population density of the prey (n_1) and the population density of the predator (n_2) can be described with the following equations:

$$dn_1/dt = n_1(b - k_1n_2)$$
$$dn_2/dt = n_2(k_2n_1 - d)$$

where b is the birth rate of the prey, d is the death rate of the predators, and k_1 and k_2 are constants. It is straightforward to write a computer program based on the Lotka-Volterra equations, computing how the population densities of the predator and prey vary with time (for example, Roberts et al. 1983).

A major difference between the Lotka-Volterra approach and the StarLogo approach is that the Lotka-Volterra equations deal with aggregate quantities (population densities), while the StarLogo program deals with the behaviors of individual creatures. Thinking in terms of individual creatures seems far more intuitive, particularly for the mathematically uninitiated. Moreover, observing the dynamics at the level of the individual creatures, rather than at the aggregate level of population densities, makes it much easier to think about and understand the population oscillations that arise.

When the StarLogo program is run, the screen is initially dominated by red turtles, with a sparse scattering of green food. Because food is scarce, many of the turtles die. But then there are fewer turtles left to eat the food, so the food becomes more dense. The few surviving turtles find themselves overwhelmed with food, and each of them rapidly increases its energy. When a turtle's energy surpasses a certain threshold, it clones, increasing the turtle population. But, of course, as the population grows too high, food again becomes scarce, and the cycle starts again.

Visually, the oscillations are striking. Red objects (turtles) and green objects (food) are always intermixed, but the density of each continually changes. Initially, the screen is overwhelmingly red, with a few green objects. As the density of red objects declines, the green objects proliferate, and the screen is soon overwhelmingly green. Then the process reverses: the density of red increases, while the density of green declines.

Depending on the particular parameters, the oscillations can take on different forms. In Benjamin's program the oscillations were damped: with each cycle, the peaks were a little less high, the troughs a little less deep. In the first cycle the turtle population dwindled to just 26 turtles, then it rose to 303 turtles. In the next cycle the population shrank to 47 turtles, then up to 244 turtles. Eventually, the turtle population stabilized between 130 and 160 turtles.

Benjamin recognized that this result depended critically on the parameters in his StarLogo program. He wondered, What would happen if the food grew just half as quickly? He figured that this new world would support fewer turtles, but how many fewer? In the original version of the `food-demon` procedure, each patch had a 1 in 1,000 chance of growing food. Benjamin changed it to 1 in 2,000.

When Benjamin ran the program, he was in for another surprise: all of the turtles died. But Benjamin, who had just finished graphing the oscillations from the previous experiment, quickly realized what had happened. "The oscillation must be between some number and negative something," he said. That is, the trough of the oscillation must drop below zero. And once the population drops below zero, it can never recover. There is no peak after a negative trough. Extinction is forever, another trapped state.

The problem lay in the initial conditions. Benjamin had started the simulation with a thousand turtles. If there were fewer initial turtles, the first trough wouldn't sink so deep. Benjamin came up with an ingenious solution. "I'll start with just one [turtle]," he explained. "It will definitely survive. I'll put money on it."

Benjamin started the program again, this time with a single turtle. For a while, the single turtle roamed the world by itself. Benjamin cheered it on: "Come on. Hang on there. Come on. Get some food." Finally, the turtle cloned, and then there were two. "He's going to live," exclaimed Benjamin.

The turtle population rose to about 130 turtles, leveled off, then fell. As before, the turtle population went up and down in a damped oscillation. Eventually, the population stabilized at about 75 turtles. So with food growing at half the rate as before, the turtle population stabilized at about half the level as before. The "equilibrium population" seemed to be proportional to the rate of food growth.

Before running the program, Benjamin had predicted that the equilibrium population would be more drastically affected by the reduction in food growth. He expected the population to stabilize with considerably fewer than 75 turtles. But after watching the program run, he developed a explanation for the proportional relationship. Looking at the dots of food on the screen, he noted that the "food density" at equilibrium looked about the same as in the previous experiment, despite the change in the rate of food growth. That made sense to him: a certain food density is needed to keep the turtles just on the brink between death and reproduction. To reach a relatively steady state, the system needed to maintain that special food density. Given that the food was growing just half as quickly as before, it made sense that the system could support only half as many turtles.

Benjamin's reasoning is an example of what Hut and Sussman (1987) dubbed "analysis by synthesis." Traditionally, synthesis and analysis have been seen in opposition to one another, two alternate ways of solving problems. But with computer-based explorations, the two approaches get mixed and blurred. It is very unlikely that Benjamin could have developed his explanation without actually viewing (and manipulating) the simulation. Only by building and creating (synthesis) was Benjamin able to develop a well-reasoned explanation for the behavior of the turtles (analysis).

New Turtle Geometry

While I was designing StarLogo, my primary goal was to develop a language for exploring biological phenomena, such as slime-mold aggregation and ant-colony foraging. But once StarLogo was up and running, I stumbled upon some unexpected ways to use the new abundance of turtles.

At one point, I typed the following commands:

```
create-turtles 5000
setxy 0 0
```

The first command created 5,000 turtles on the screen (in random positions, with random headings). The second command made all of the turtles move to the middle of the screen, to the Cartesian point (0,0). Only a single turtle was visible on the screen. But in fact, that single turtle was at the top of a very tall pile of turtles, with 4,999 turtles underneath it—somewhat like the pile of turtles in Dr. Seuss's *Yertle the Turtle*. (Note: StarLogo turtles are created, by default, with their "pens" up. So, in the second command, they do *not* draw lines as they move to the point (0,0).)

I grinned. The pile of turtles was a neat trick. Then, suddenly, I realized an even better trick. I typed this command:

```
forward 50
```

The 5,000 turtles exploded outward from the center of the screen. Since StarLogo, by default, creates turtles with random headings, the turtles all moved in random directions. But their overall pattern was anything but random. After the turtles had moved forward five steps, they formed a circle of radius 5 (centered on the point (0,0)). After they had moved another five steps, they formed a circle of radius 10. So the overall effect was an expanding circle (always centered on the point (0,0)). The circle grew until it reached a radius of 50.

This trick works only if there are *lots* of turtles. If there were only 500 turtles (instead of 5,000), the circle of turtles would have lots of holes in it. All of the turtles would lie on the same circle, but they wouldn't appear as a complete circle. To give the appearance of a complete circle, turtles must be distributed around the entire circumference of the circle. That could be done explicitly, by giving each turtle a different heading:

```
clear-all
create-turtle 360
setxy 0 0
set-heading who
forward 50
```

In this case each turtle sets its heading to be equal to its unique ID number (reported by the StarLogo primitive who). With 360 turtles the ID numbers range from 0 to 359, so the turtles' headings are evenly distributed around the circle, one at every integer heading. As the turtles move forward, they fill in the entire circumference of the circle. (Of course, if the circle became large enough, holes would appear. More turtles, with headings distributed at finer resolution, would be needed. The battle is never ending: for ever-larger circles, there is a need for ever-more turtles, with ever-finer resolution of headings.)

The approach with 360 turtles, with headings evenly distributed, works as desired. But the original approach with 5,000 turtles has a nicer feel to it (at least for me). It makes use of the law of large numbers. Each of the 5,000 turtles has a random heading. There is a chance, of course, that all 5,000 turtles could have headings between 0 and 90, so they would form only a quarter-circle. But statistically, the 5,000 turtles are almost certain to fill in the entire circumference just as well as the 360 carefully arranged turtles. This approach might seem wasteful of turtles. Why use 5,000 turtles when 360 will do? But the approach with 5,000 turtles has a different aesthetic to it. It allows order to arise not from top-down planning but from the statistical properties of a random distribution.

This approach can be generalized into a circle procedure, which creates a circle of any radius r, centered at any point (x,y):

Observer Procedures

```
to circle x y r
clear-all
create-turtle 5000
setxy x y
forward r
end
```

This approach to drawing a circle represents a new form of turtle geometry. In traditional turtle geometry, the Logo turtle uses a "pen" to draw various geometric shapes and patterns (Abelson and diSessa 1980). For example, the following command makes the Logo turtle draw a circle:

```
repeat 360 [forward 1 right 1]
```

The turtle takes a step forward, then turns a degree to the right, then another step forward, and so on. After 360 steps, the turtle returns to its starting point, having completed a circle. (Actually, it draws a regular polygon with 360 sides. But if you increase the number of steps, and decrease the turning angle, the polygon becomes closer and closer to a circle, approaches a circle as its limit.)

The StarLogo form of turtle geometry, while still based on the Logo turtle, is quite different. Rather than a single turtle drawing geometric shapes and patterns, a collection of turtles use their own "bodies" to form geometric shapes and patterns. Rather than turtles *drawing* circles, the turtles *are* the circle.

This new StarLogo turtle geometry is not limited to circles. Consider what happens if you type the following:

```
clear-all
nowrap
create-turtles 5000
sety xpos
```

The third command fills the screen with turtles in random positions. The fourth command tells each turtle to calculate its x-position, then set its y-position to have the same value. Visually, the result is striking: the random mess of creatures transforms itself into a diagonal line, from the bottom-left corner of the screen to the upper-right. It is the line $y=x$. (As in the first circle example, this example uses lots of turtles to make sure that there are no holes in the line. The same effect is possible with fewer turtles, if the turtles are distributed evenly across the screen.)

You can use this same approach to graph any function. If you type the command

```
sety xpos * xpos
```

the turtles align themselves into the parabola $y=x^2$. And if you type the command:

```
sety 50 * sin xpos
```

the turtles align themselves into a sine wave (with an amplitude of 50).

There are several different ways to think about mathematical functions. You can think about a function as an "input-output machine": a function takes an input value and returns a unique output value. Or you can think about a function in terms of its graphical representation, which shows all possible input-output combinations. Students learning about functions often have trouble understanding the connection between these two representations. The new StarLogo approach to graphing could help make that connection clearer. Each turtle acts as a simple input-output machine, using its x-position as the input to the machine, and adjusting its y-position to indicate the output value of the machine. But taken together, the collection of all turtles forms a graphical representation of the function.

Patch Geometry?

StarLogo offers yet another approach for thinking about graphing. In this approach, you need to focus on the patches rather than the turtles. Consider what happens if you type the following command:

```
if patch-ypos = patch-xpos [set-patchcolor green]
```

Each patch asks itself, "Does my x-position equal my y-position? If so, I should turn green." Every patch on the line $y=x$ turns green. You can use this approach to graph any function (and even non-functional relations).

These StarLogo geometry examples seem somewhat different from the earlier StarLogo explorations (such as the slime-mold aggregation and the traffic-jam formation). The reason is that there is no interaction between the turtles (or patches) in the geometry examples. Each of the turtles (or patches) just does its own thing. The whole is precisely the sum of the parts. As a result, the geometry examples seem less compelling than the earlier StarLogo explorations. Parallelism without interaction loses what is most exciting and intriguing about parallelism.

Can we construct StarLogo geometry programs that *do* involve interaction, in which the final geometric forms emerge from interactions among the turtles? Indeed we can. Here's an alternative approach for making a circle with StarLogo turtles. Imagine a bunch of "circle-turtles" trying to maintain a fixed distance from a special "center-turtle." If a circle-turtle is too close to the center-turtle, it moves away a little. If it is too far away, it moves a little closer. So the circle-turtles should form a circle around the center-turtle, right? Not quite. What if all of the circle-turtles start in a vertical line directly "north" of the center-turtle? Then they will all drift to the exact same point. The circle-turtles form a point, not a circle. We need some way to spread the circle-turtles around the center-turtle.

One way to do that is to add another rule: each of the circle-turtles should repel the other circle-turtles a little bit. That will force the circle-turtles to spread out as much as possible. They should reach an equilibrium when they are evenly spaced around the center-turtle. This approach (suggested to me by Brian Silverman) is an example of a general technique that might be called "constraint and noise." The first rule (maintain a fixed distance from the center-turtle) is the constraint. The second rule (repel the other circle-turtles) provides the noise that is needed to spread the turtles evenly around the circle.

Here is a StarLogo implementation of this strategy.

Observer Procedures

```
to setup
clear-all
nowrap
create-turtle 501
turtle-setup
end
```

Turtle Procedures

```
to turtle-setup
turtles-have radius
set-radius 20
if who = 500
   [set-color green]
if who < 500
   [activate-demon [constraint-demon
                    noise-demon]]
end

to constraint-demon
local [error center-x center-y]
set-center-x ask 500 [xpos]
set-center-y ask 500 [ypos]
set-heading towards center-x center-y
set-error (distance center-x center-y) - radius
forward 0.1 * error
end

to noise-demon
local turtle-to-avoid
set-turtle-to-avoid random 500
set-heading towards ask turtle-to-avoid [xpos]
                    ask turtle-to-avoid [ypos]
back 0.2
end
```

Programming Notes

Note that the constraint-demon and noise-demon are activated selectively, only for the circle-turtles, not for the center-turtle.

In the noise-demon, the circle-turtles do not actually repel *all* other circle-turtles. Rather, on each time step, each circle-turtle randomly chooses one other circle-turtle and moves slightly away from it. Over a long time, this strategy has roughly the same effect as repelling all other circle-turtles.

When this program is run, the circle-turtles spread themselves around the green center-turtle, forming a circle with a radius of 20. While the demons are running, we can make certain changes and observe how the turtles react. For example, if we type

```
set-radius 30
```

the circle-turtles will all move away from the green center-turtle, gradually converging to a circle with a radius of 30. If we type

```
if color = green [set-heading 0 forward 50]
```

the green center-turtle will move due "north," jumping outside of the circle. The circle-turtles will follow to the north, pursuing the center-turtle. At first, the circle-turtles will form an arc to the south of the center-turtle. But, with time, the noise-demon will force the circle-turtles to spread out, forming a complete circle around the center-turtle.

This program uses a privileged turtle (the center-turtle) that acts differently from all of the others. But with a few small changes it is possible to write a constraint-and-noise program that creates a circle without any privileged turtles. Imagine that each circle-turtle, instead of trying to maintain a fixed distance from a special center-turtle, tries to maintain a fixed distance from one other "target" circle-turtle. Each circle-turtle has a different target: for example, circle-turtle 1 might try to maintain a fixed distance from circle-turtle 2, while circle-turtle 2 tries to maintain a fixed distance from circle-turtle 3, and so on.

What will happen? The circle-turtles arrange themselves in a polygon with roughly equal sides. But the polygon might have lots of concavities. With lots of turtles it looks like a jumbled mess. But if we again add a noise-demon, the turtles will push away from one another until the concavities are gone. The result is a fully convex regular polygon. With enough turtles, the polygon resembles a circle.

Below is a StarLogo implementation of this strategy.

Observer Procedures

```
to setup
clear-all
nowrap
create-turtle 50
turtle-setup
end
```

Turtle Procedures

```
to turtle-setup
turtles-have [target target-distance]
set-target remainder who + 1 50
set-target-distance 5
activate-demon [constraint-demon noise-demon]
end

to constraint-demon
local [target-x target-y error]
set-target-x ask target [xpos]
set-target-y ask target [ypos]
set-heading towards target-x target-y
set-error (distance target-x target-y)
              - target-distance
forward 0.5 * error
end

to noise-demon
local turtle-to-avoid
set-turtle-to-avoid random 50
set-heading towards ask turtle-to-avoid [xpos]
                    ask turtle-to-avoid [ypos]
back 0.2
end
```

Programming Notes

In this case there is no special center-turtle, so the constraint-demon and noise-demon are activated for all of the turtles.

I used relatively few circle-turtles in this example (50), because with lots of circle-turtles, it takes a long time for all of the concavities to disappear.

The noise-demon is exactly the same as in the previous example, except that it uses fewer turtles.

Is this new form of turtle geometry better than traditional turtle geometry? That is the wrong question to ask. The point is not to provide *better* ways of doing geometry but to provide *more* ways of doing (and thinking about) geometry.

There are (at least) two major reasons for developing new ways of doing geometry. First, different people find different approaches more accessible. Some people might find the traditional Euclidean approach intuitive and accessible, others might prefer turtle geometry, still others might connect most easily with the new StarLogo turtle geometry. Too often, schools give special status to particular ways of thinking about mathematical and scientific ideas. By privileging certain types of thinking, they exclude certain types of thinkers.

Second, everyone can benefit from learning *multiple ways of thinking* about things. Understanding something in just one way is a rather fragile kind of understanding. Marvin Minsky has said that you need to understand something at least two different ways in order to really understand it. Each way of thinking about something strengthens and deepens each of the other ways of thinking about it. Understanding something in several different ways produces an overall understanding that is richer and of a different nature than any one way of understanding. Thus the new StarLogo turtle geometry has the potential to supplement and reinforce all of the other ways of thinking about geometry.

Forest Fire

A fire starts on the edge of a forest. What is the chance that the fire will spread all the way through the forest?

In certain extreme situations the behavior of the fire is easy to predict. If the forest is very densely populated with trees, the fire is likely to spread. And if the forest is only sparsely populated, the fire is likely to die out. (In the limiting case of no trees at all, the fire will not get anywhere!) But what happens for the in-between densities of trees? Is there a critical density that is needed for the fire to propagate?

Questions like these are studied in a branch of mathematics known as percolation theory. Percolation problems typically involve two intermingled "substances" with different properties. In the case of the forest fire, the forest is composed of trees (which support the spread of the fire) and empty space (which inhibits the spread of the fire). Which side wins? The trees or the empty space?

Many other situations can be described in a similar way (Peterson 1988). Consider oil seeping through porous rock. The porous rock is composed of the rock itself (which inhibits the spread of the oil) and empty space (which supports the spread of the oil). If there is enough empty space (a "high density" of empty space), oil will spread large distances through the rock—just as the forest fire spreads through the trees. (Notice that empty space plays opposite roles in these two cases: it inhibits propagation of the fire but supports propagation of the oil.)

New high-temperature superconductors are also based on a percolation process. These superconductors are actually mixtures of superconducting material and resistive material. The superconducting material lets electrons move through freely, while the resistive material inhibits electrons. If there is enough superconducting material in the mixture, electrons are able to spread—and the mixed substance acts like a superconductor.

I decided to explore percolation phenomena using the StarLogo patches. In the program I turn some of the patches into "trees" and leave the rest as "empty space." Then I start a fire at the left edge of the screen. Each tree follows a simple rule: if it catches fire, it spreads the fire to any neighboring trees to its north, south, east, or west.

Observer Procedures

```
to setup percentage
clear-all
patch-setup percentage
end
```

Patch Procedures

```
to patch-setup percentage
setup-trees percentage
setup-fire
setup-border
activate-demon [spread-demon burn-demon]
end
```

```
to spread-demon
if on-fire?
    [ask patch 0    [if patchcolor = green
                        [set-patchcolor red]]
      ask patch 90   [if patchcolor = green
                        [set-patchcolor red]]
      ask patch 180 [if patchcolor = green
                        [set-patchcolor red]]
      ask patch 270 [if patchcolor = green
                        [set-patchcolor red]]]
end

to burn-demon
if on-fire? [set-patchcolor patchcolor - 1]
end

to on-fire?
(patchcolor >= 1) and (patchcolor <= 5)
end

to setup-trees percentage
if (10 * percentage) > (random 1000)
      [set-patchcolor green]
end

to setup-fire
if patch-xpos = (left-edge + 1)
      [set-patchcolor red]
end

to setup-border
if (patch-xpos = left-edge)
      or (patch-xpos = right-edge)
      or (patch-ypos = top-edge)
      or (patch-ypos = bottom-edge)
   [set-patchcolor blue]
end
```

Programming Notes

The burn-demon procedure is included for visual effect. When a tree catches fire, its color is set to bright red. Then the burn-demon gradually dims the intensity of the color, giving the visual effect of a fire dying out.

When the program starts, there is a neat line of red fire on the left edge of the screen. Then the fire spreads from tree to tree. In some places the fire quickly reaches a dead end, surrounded by empty space. In other places the fire continues to spread from tree to tree. Overall, the fire paths form fractal-like patterns on the screen (figures 3.9 and 3.10).

The focus of my investigation: Under what conditions will the fire spread all the way to the right edge of the screen? For that to happen, there has to be a connected path of trees all the way from the left edge to the right edge. The path needn't be straight or direct: it could wander all over the screen. But the path must be connected all the way from left to right.

I ran the program many times, with different densities of trees (using different inputs to the setup procedure). As expected, when the tree density is near 100 percent, the fire spreads quickly and easily, always reaching the right edge. When the tree density is near 0 percent, the fire quickly dies out.

At in-between densities, the behavior is less intuitive. What happens when the tree density is 20 percent (that is, when 20 percent of the patches are trees)? The fire dies out quite quickly, never reaching the right edge of the screen. How about 30 percent trees? The fire tends to spread a little further, but not much. It never spreads all the way across the screen. How about 40 percent trees? Or 50 percent? The result is still the same: the fire always dies out. Even at 55 percent trees, the result is the same (figure 3.9). I ran the program 100 times with a tree density of 55 percent, and the fire never spread all the way across the forest. Not even once.

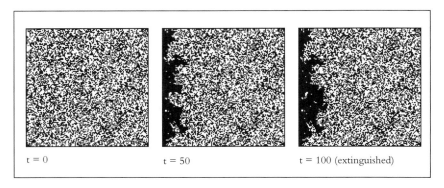

t = 0   t = 50   t = 100 (extinguished)

Figure 3.9
Fire in forest with 55% tree density (trees in white, fire starts at left edge)

But as the tree density increases above 55 percent, the results change very quickly. At a tree density of 59 percent, the fire spread across the forest about half of the time in my trials. At a tree density of 63 percent, the fire spread across the forest every time I tried it (figure 3.10). I ran the program 100 times with a tree density of 63 percent, and the fire spread across the forest all 100 times. (Of course, there is *some* chance that the fire *could* fail to spread across such a forest, but the chances of such a failure are minuscule.)

So there is a sharp transition between 55 percent and 63 percent (see figure 3.11). The transition becomes even sharper when the size of the forest is increased from 128 × 128 to 512 × 512 (see figure 3.12). In fact, percolation theory predicts a critical density around 59.2 percent. If the forest were infinitely large, the transition would occur precisely at critical density. At tree densities less than critical density, the fire would never propagate. At tree densities greater than critical density, the fire would always propagate.

The idea of critical thresholds comes up quite often in explorations of decentralized systems. Often, small changes in density (or some other

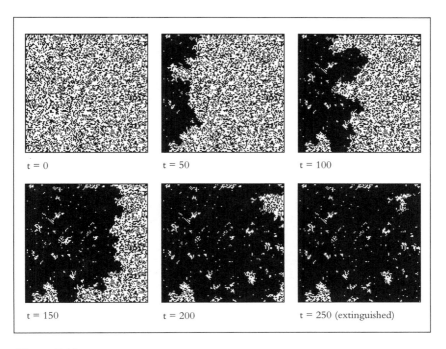

t = 0   t = 50   t = 100

t = 150   t = 200   t = 250 (extinguished)

Figure 3.10
Fire in forest with 63% tree density (trees in white, fire starts at left edge)

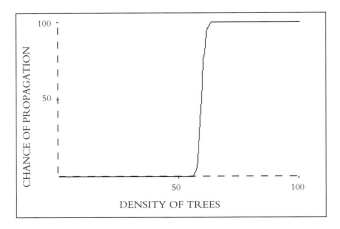

Figure 3.11
Sharp transition in forest-fire behavior

DENSITY OF TREES	CHANCE OF PROPAGATION (100 TRIALS)	
	128 x 128 forest	512 x 512 forest
55.0	0	0
55.5	0	0
56.0	1	0
56.5	3	0
57.0	6	1
57.5	10	1
58.0	19	5
58.5	36	21
59.0	50	60
59.5	61	84
60.0	76	98
60.5	80	100
61.0	93	100
61.5	96	100
62.0	96	100
62.5	98	100
63.0	100	100

Figure 3.12
The larger the forest, the sharper the transition

property) can lead to significant qualitative changes for the overall system. In the StarLogo slime-mold program there is a critical threshold in the density of slime-mold cells. When the cell density is below the critical threshold, the cells do not aggregate into clusters. Above the threshold, clusters begin to form. Similarly, in the ant-foraging program a critical density of ants is needed to sustain a pheromone trail between food and nest.

So when I ran the forest-fire program, I was not too surprised by the existence of a critical threshold in the tree density. But as I watched the program running, something started to bother me. I was reminded of a game that I had played as a child. The game, called Twixt, had a square board with two blue edges (across from one another) and two red edges (also across from one another). One player tried to build a bridge between the two blue edges, while the other player tried to build a bridge between the two red edges. The two players alternated turns. There was a neat symmetry in the game. Offense and defense were really the same. Only one player could successfully build a bridge. Once one player completed a bridge across the game board, the other player was effectively blocked. It was a zero-sum game, always one winner and one loser.

I began to wonder: Isn't the forest-fire problem just a modified version of Twixt? The fire "wins" (spreads across the entire forest) if there is a "bridge" of trees from the left edge to the right edge. And the fire "loses" if there is a bridge of empty spaces from the top edge to the bottom edge. Isn't the problem symmetric, like Twixt? If so, why isn't the critical density at 50 percent (rather than 59 percent)? Why do the empty spaces have an apparent advantage in blocking the fire?

This problem bothered me for a while. But then I realized the flaw in my logic. The bridges formed by the trees can move only north, south, east, or west. The bridges formed by the empty spaces can also move along the diagonals. In other words, a diagonal bridge of empty spaces effectively blocks the fire from spreading, but the fire cannot spread along a diagonal bridge of trees. So there is an asymmetry in the problem. The empty spaces *do* have an advantage in this game. So if the program is set up with 50 percent trees and 50 percent empty space, the empty space "wins" and the fire does not spread.

To test out this idea, I changed the StarLogo program so that the fire could spread in any of eight directions (instead of just four). With this change, the advantage was reversed: now the fire could spread along diagonal bridges of trees, and it could no longer be blocked by diagonal bridges of empty spaces. And, sure enough, the critical density changed accordingly. In the original StarLogo program, the trees needed a 59 percent density to propagate the fire (and the empty spaces needed only a 41 percent density to block it). In the revised program, the trees needed only a 41 percent density to propagate the fire (and the empty spaces needed a 59 percent density to block it).

Recursive Trees

On certain packages of Morton's salt, there is a picture of a little girl carrying a package of salt, which has a picture of a little girl carrying a package of salt, which has a picture . . .

Almost everyone who has seen this picture remembers it. The picture seems to evoke a strong reaction. People find the picture amusing—and, perhaps, a bit unsettling. The chain of ever-smaller girls can't go on forever—or can it? People have a similar reaction to the joke about the person who, when granted three wishes by a genie, uses the third wish to wish for three more wishes. That wouldn't really work—would it?

The Morton's salt package and the genie joke are classic examples of *recursion*. Examples of recursion arise in a wide variety of situations and domains. As Hofstadter (1979) writes, "The concept is very general. (Stories inside stories, movies inside movies, paintings inside paintings, Russian dolls inside Russian dolls (even parenthetical comments inside parenthetical comments!)—these are just a few of the charms of recursion.)"

The charms and excitement of recursion are certainly not lost on children. In stores that happen to have two mirrors set up directly across from one another, children love to position themselves so that they can see an endless chain of ever-smaller reflections of themselves. And in writing Logo programs, children quickly recognize recursive procedures (that is, procedures that call versions of themselves as subprocedures) as a source of great power. Papert (1980) notes, "Of all ideas I have introduced to children, recursion stands out as the one idea that is particularly able to evoke an excited response."

The tree has become a symbol of Logo recursion. You can think of a tree as a trunk supporting two smaller trees, each of which is a trunk supporting two smaller trees, and so on. To draw such a tree in Logo, you can write a procedure like this:

```
to tree length
if length < 2 [stop]
forward length
left 45
tree length * 0.7
right 90
tree length * 0.7
left 45
back length
end
```

The tree procedure first draws the "trunk" of the tree (using forward), then it calls the tree procedure twice more to draw two smaller trees (one pointed 45 degrees to the left, the other pointed 45 degrees to the right) at the end of the trunk. Each subtree is 70 percent as large as the "master" tree that it is part of. Were it not for the first line of the procedure (if length < 2 [stop]), the recursion would go on forever, trying to draw smaller and smaller trees. The first line acts as a "stop rule" to prevent infinite regress: the recursion stops whenever a subtree has a length of less than 2.

Like many recursive procedures, the tree procedure seems almost magical. A mere eight lines of Logo code produce a wonderfully detailed tree—and in two of the eight lines, the procedure calls itself! As written, the procedure draws a perfectly symmetric tree. To make the tree look more like a "real" tree, with some asymmetry in the branch lengths and angles, you would need to make just a few minor changes (using the Logo random procedure).

But there is a problem with the tree procedure: it is quite difficult to understand. For example, why does the turtle need to turn left and go backward at the end of the procedure? Isn't that wasted effort? After the turtle draws the two smaller trees at the appropriate angles, shouldn't it be done? What's the point of turning again if the procedure is over?

In fact, the final left turn and backward movement are needed. When the tree procedure calls itself recursively to make a smaller subtree, it assumes that the turtle (after drawing the subtree) will end up where it started. So at the end of the tree procedure, the turtle must turn left and go backward, to return to its starting point. Even if these final movements aren't necessary for the "top-level" call of the tree procedure, they are necessary for each of the recursive calls. So they must be included in the tree procedure.

The idea of a "state-preserving procedure" (in which the turtle returns to its initial position and its initial heading) is a powerful idea, with applicability in many other situations. But it is also a difficult idea. Many people, when writing the recursive procedure to draw a tree, forget to return the turtle to its initial state, leading to a very strange-looking tree.

There's another problem with the `tree` procedure. When the turtle draws the tree, it draws it in a strange way (figure 3.13). First it draws all of the leftmost branches of the tree, and then it works its way over until it draws all of the rightmost branches. (And if you forget to include the stop rule, the turtle will draw *only* the leftmost branches.) That's certainly not the way most people *think about* the tree when writing the `tree` procedure. Most people think about the turtle drawing the tree from bottom to top, more like the way a real tree grows. First, the turtle should draw the trunk, then the next two branches, then the four sub-branches that come off of those branches, and so on. Thus there is a significant mismatch between people's expectations of how the tree should "grow," and how the turtle actually draws the tree. This mismatch is a major source of confusion.

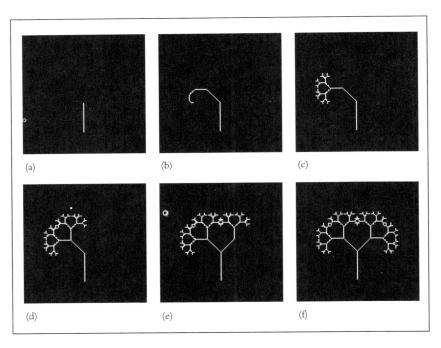

Figure 3.13
Drawing a recursive tree with traditional Logo

So there are two problems: the "state-preservation" problem and the "unnatural dynamics" problems. Is there a way to draw a tree that gets around these problems? Not if you have just one turtle. But if you have lots of turtles and all of them can draw in parallel (as in StarLogo), another approach is possible.

Let's start with a single turtle, heading "north" on the screen. The turtle draws the trunk of the tree, then "clones" two new turtles, one heading 45 degrees to the left and another heading 45 degrees to the right. That's all for the initial turtle. Its work is done, and it dies. Then we call the tree procedure recursively. But this time there are two turtles (cloned from the initial turtle). The two new turtles draw (smaller) branches, then each of them clones two new turtles (for a total of four new turtles), and they die. On the next recursive call, there are eight turtles, then sixteen, and so on (figure 3.14). The number of turtles grows exponentially on each recursive call, as the turtles draw smaller and smaller subbranches. As before, we need a stop rule to stop the procedure as the branches get very small.

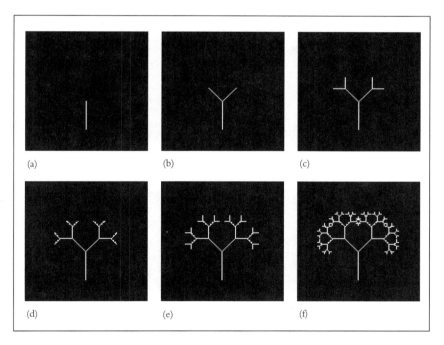

Figure 3.14
Drawing a recursive tree with StarLogo

In effect, this StarLogo approach trades off turtles for procedure calls. Whereas the original Logo `tree` procedure uses *multiple recursive calls* to draw multiple subtrees, the StarLogo approach uses *multiple parallel turtles* to draw the multiple subtrees. Below is an implementation of this strategy.

With this approach, the tree is drawn in a more "natural" way—more like the way real trees grow, and more like the way people are likely to think about trees growing. First the big trunk, then two branches off the main trunk, then four subbranches, and so on. All of the branches at the same level of the tree are drawn at the same time. Also, there is no need for any turtle to backtrack after it is finished its work. Once a turtle has drawn its branch, it just creates two new turtles, then dies peacefully.

Observer Procedures

```
to tree length
if length < 2 [stop]
draw-branch length
tree length * 0.7
end

to setup
clear-all
create-turtle 1
turtle-setup
end
```

Turtle Procedures

```
to turtle-setup
setxy 0 -60
set-heading 0
pendown
end

to draw-branch length
forward length
clone [left 45]
clone [right 45]
die
end
```

Programming Notes

These procedures look rather straightforward, but there is some hidden subtlety. The `draw-branch` procedure includes two successive `clone` commands. The trickiness arises with the second `clone` command. We want only the original turtles to clone, not the new turtles that were just created by the first `clone` command. Similarly, with the `die` command, we want only the original turtles to die, not the turtles that were just created by the two `clone` commands.

StarLogo does the "right thing" in this case. The command `draw-branch length` means that each *currently existing* turtle should execute the `draw-branch` procedure. Newly cloned turtles do nothing within this procedure.

Although this approach works, it feels very centralized. The observer acts like a dictator, telling the turtles what to do at every step. The observer even keeps track of how far the turtles should move on each step. I wanted to create a greater sense of the turtles acting on their own. So I added a turtle demon. Each turtle runs the demon repeatedly, without any direct influence from the observer. For the turtles to run "by themselves" in this way, each turtle needs to know how far it should move. In other words, each turtle needs a "length-gene," indicating the length of its branch.

With this approach, it is possible to generate a much greater variety of tree shapes, since turtles on the same level of the tree no longer need to go the same distance. Different turtles can have different values for the length-gene. If the turtles on the right side of the tree have length-genes with larger values than turtles on the left, the tree will become asymmetric, with longer branches on the right. Here is one way to implement this idea.

Observer Procedures

```
to setup length
clear-all
create-turtle 1
turtle-setup length
end
```

Turtle Procedures

```
to turtle-setup length
turtles-have length-gene
setxy 0 -60
set-heading 0
set-length-gene length
pendown
activate-demon draw-branch-demon
end

to draw-branch-demon
ifelse length-gene < 2
   [die]
   [forward length-gene
    clone [mutate-length-gene left 45]
    clone [mutate-length-gene right 45]
    die]
end

to mutate-length-gene
set-length-gene length-gene * shrink-factor
end

to shrink-factor
0.5 + (0.1 * random 5)
end
```

This program has only one demon procedure: draw-branch-demon. Each turtle runs this procedure over and over. On each execution of the demon, each turtle moves forward a distance based on its length-gene variable. Then it gives birth to two new turtles, and it dies. Each of the "children" has a slightly mutated version of the parent's length-gene. Each new turtle gets a smaller value of length-gene than its parent, so each generation draws slightly shorter branches. The mutation includes some randomness, so the children of a given turtle are likely to draw branches differing slightly in length from one another. These differences are likely to persist as the tree grows, since the length-gene of a new turtle is based (in part) on the length-gene of its parent. As a result, long tree branches are likely to have relatively long subbranches attached to them.

When a new turtle is born with a length-gene less than 2, it dies before it clones any new turtles. Eventually all of the turtles die, and the tree is complete.

Looking Ahead: From Explorations to Reflections

These nine StarLogo explorations involve many different sorts of things: termites and forest fires, ant colonies and automobiles, slime mold and frogs. It wouldn't be too surprising to see StarLogo programs involving cabbages and kings—or walruses and carpenters.

There is a point to this diversity. Ideas about decentralization and self-organization are not restricted to any particular domain. They arise in many different fields of study. Next, in Reflections, I discuss some unifying ideas about decentralized systems—and some unifying ideas about how people think about decentralized systems.

4

Reflections

A man's spirit must take a different shape
if he believes that all sorting in the universe
is due to an external agent.

—Gregory Bateson, *Steps to an Ecology of Mind*

The Centralized Mindset

One day, shortly after I developed the first working prototype of
StarLogo, Marvin Minsky wandered into my office. On the computer
screen he saw an early version of my StarLogo slime-mold program.
There were several green blobs on the screen (representing a chemical
pheromone), with a cluster of turtles moving around inside each blob. A
few turtles wandered randomly in the empty space between the blobs.
Whenever one of these turtles wandered close enough to a blob, it moved
toward the blob and joined the cluster of turtles inside.

Minsky, one of the founders of the field of artificial intelligence, asked
me what I was working on. I explained that I was experimenting with
some self-organizing systems. Minsky looked at the screen for a while,
then said, "But those creatures aren't self-organizing. They're just moving
toward the green food."

Minsky had assumed that the green blobs were pieces of food, placed
throughout the turtles' world. In fact, the green blobs were created by the
turtles themselves. Each turtle dropped green pheromone behind itself as
it moved, while also "sniffing" ahead to try to follow the pheromone
scent. But Minsky didn't see it that way. Rather than seeing creatures
organizing themselves, he saw the creatures organized around some preexisting pieces of food. He assumed that the pattern of aggregation was
determined by the placement of food. And he stuck with that interpretation (at least for a while) even after I told him that the program involved
self-organization.

Minsky has thought more—and more deeply—about self-organization and decentralized systems than almost anyone else. When I explained the rules underlying the slime-mold program to him, he understood immediately what was happening. But Minsky's initial assumption was revealing. When he first saw the simulation on the computer screen, he resisted my suggestion that the simulation was based on self-organization. Instead, he assumed that the pattern was determined by a more centralized cause (preexisting pieces of food). The fact that *even Marvin Minsky* had this reaction is an indication of the powerful attraction of centralized explanations.

This inclination toward centralization was apparent in many StarLogo projects. When trying to make sense of StarLogo programs, people tended to assume centralized causes. And when trying to construct StarLogo programs, they often imposed centralized control. There was evidence of the centralized mindset in all types of StarLogo users, from the scientifically sophisticated to the scientifically naive, from expert hackers to novice programmers. Some people were more successful than others in moving beyond the centralized mindset. But everyone slipped into centralized thinking at times.

People's inclination for centralization in StarLogo projects provides a glimpse at a much more general phenomenon. People seem to have a strong preference for centralization in almost everything they think and do. People tend to look for *the* cause, *the* reason, *the* driving force, *the* deciding factor. When people observe patterns and structures in the world (for example, the flocking patterns of birds or the foraging patterns of ants), they often assume centralized causes where none exist. And when people try to create patterns and structures in the world (for example, new organizations or new machines), they often impose centralized control where none is needed.

In some ways, the pervasiveness of the centralized mindset might seem surprising. After all, aren't we living in an Era of Decentralization? Many different sorts of things—organizations, technologies, scientific models— are all becoming decentralized. Ideas about decentralization are rapidly spreading through our culture. Magazines and journals are filled with articles related to decentralization. If that is the case, isn't it surprising that people still view the world in such a centralized way?

Actually, it isn't so surprising if we look at the growing interest in decentralization from a different perspective: Why are people becoming more interested in decentralized ideas *now*? Why didn't it happen before? Why have people resisted decentralized approaches in the past? What underlies this persistence of resistance? What made people cling onto centralized approaches so tightly for so long?

Why, for example, did scientists assume for so long that bird flocks must have leaders? Why did Freudian psychologists focus for so long on an "ego-centered" model that puts one agent at center stage? Why did computer scientists focus for so long on centralized computer architectures? People seem to have very strong attachments to centralized ways of thinking.

The centralized mindset is particularly apparent in the history of biology. Until the mid-nineteenth century, almost everyone embraced the idea that living systems were designed by some God-like entity. Even scientists were convinced by the so-called watchmaker argument (or the argument from design) proposed by theologian William Paley in his 1802 book *Natural Theology*. Paley noted that watches are very complex and precise objects. If you found a watch on the ground, you could not possibly believe that such a complex object had been created by random chance. Instead, you would naturally conclude "that the watch must have had a maker: that there must have existed, at some time, and at some place or other, an artificer or artificers, who formed it for the purpose which we find it actually to answer; who comprehended its construction, and designed its use" (Paley, 1802, quoted in Dawkins 1986).

For Paley, the same logic applies to living systems: "Every indication of contrivance, every manifestation of design, which existed in the watch, exists in the works of nature; with the difference, on the side of nature, of being greater or more, and that in a degree which exceeds all computation." So living systems, like watches, must have a maker, concluded Paley. It is not surprising that scientists accepted Paley's argument in the early nineteenth century, since there were no viable alternative explanations for the complexity of living systems. What *is* surprising is how strongly scientists held onto centralized beliefs even after Darwin provided a viable (and more decentralized) alternative. Science historian Ernst Mayr (1982) notes that biologists put up "enormous resistance" to Darwin's theories for a full 80 years after publication of *Origin of Species,* generally preferring more centralized alternatives.

Indeed, the history of evolutionary biology is filled with examples of scientists remaining committed to centralized explanations, even in the face of discrediting evidence. When fossil records showed that very different creatures existed at different times in history, scientists did not give up on ideas of supernatural creation. Rather, they hypothesized that there must have been a whole series of extinctions and new creations. In the twentieth century, as the genetic basis of evolution became understood, scientists initially adopted a too-centralized view of genes, focusing on the actions and fitness values of individual genes, rather than studying interactions among genes.

Even today, centralized thinking persists in evolutionary debates. In trying to explain the periodic massive extinctions of life on Earth, many scientists assume some external cause—for example, periodic waves of meteors hitting the Earth. But more decentralized explanations are possible. Recent computer simulations show that simple interactions within the standard evolutionary process can give rise to periodic massive extinctions, without any outside intervention (Lindgren 1991).

The history of research on slime-mold cells, as told by Evelyn Fox Keller (1985), provides another example of centralized thinking. As discussed earlier, slime-mold cells sometimes gather together into clusters. For many years, scientists believed that the aggregation process was coordinated by specialized slime-mold cells, known as "founder" or "pacemaker" cells. According to this theory, each pacemaker cell sends out a chemical signal, telling other slime-mold cells to gather around it, resulting in a cluster. In 1970, Keller and Segel (1970) proposed an alternative model, showing how slime-mold cells can aggregate without any specialized cells. Nevertheless, for the following decade other researchers continued to assume that special pacemaker cells were required to initiate the aggregation process. As Keller (1985) writes, with an air of disbelief, "The pacemaker view was embraced with a degree of enthusiasm that suggests that this question was in some sense foreclosed." By the early 1980s, based on further research by Cohen and Hagan (1981), researchers began to accept the idea of aggregation among homogeneous cells, without any pacemaker. But the decade-long resistance serves as some indication of the strength of the centralized mindset.

The centralized mindset has undoubtedly affected many theories and trends in the history of science. Just as children assimilate new information by fitting it into their preexisting models and conceptions of the world, so do scientists. As Keller puts it, "In our zealous desire for familiar models of explanation, we risk not noticing the discrepancies between our own predispositions and the range of possibilities inherent in natural phenomena. In short we risk imposing on nature the very stories we like to hear." In particular, we risk imposing centralized models on a decentralized world.

Even today, in the midst of the Era of Decentralization, most people seem inclined to view the world in a centralized way. People continue to construct centralized theories to explain the patterns they see in the world. In trying to understand the origin of the species, for example, many people still resist the idea of evolution by natural selection. More than a century after Darwin, many people continue to believe that only a centralized Designer of Life could have created the wonderful diversity and complexity of the living world.

Conspiracy theories are another example of centralized thinking. For almost every perceived problem in society, people look for a clearly identifiable culprit to blame. Something is wrong with the world economy? Blame the Trilateral Commission. Traditional family values are on the decline? Blame the producers in Hollywood. In general, people tend to focus blame on a centralized cause rather than sort through the complex, interacting factors that underlie most social phenomena.

People also view the workings of the economy in centralized ways, assuming singular causes for complex phenomena. Children, in particular, seem to assume strong governmental control over the economy. (Of course, governments *do* play a large role in most economies, but children assume that governments play an even larger role than they actually do.) In interviews with Israeli children between 8 and 15 years old, psychologist David Leiser (1983) found that nearly half of the children assumed that the government sets all prices and pays all salaries. Even children who said that employers pay salaries often believed that the government provides the money for the salaries. A significant majority of the students assumed that the government pays the increased salaries after a strike. And many younger children had the seemingly contradictory belief that the government is also responsible for organizing strikes. As Leiser writes, "The child finds it easier to refer unexplained phenomena to the deliberate actions of a clearly defined entity, such as the government, than to impersonal 'market forces.'"

By Lead or by Seed

The centralized mindset can manifest itself in many different ways. When people observe patterns or structures in the world, they sometimes assume that a *leader* orchestrated the pattern. For example, when people see patterns in an ant colony, they often assume that the queen ant orchestrated the pattern. And when people see patterns in human society, they often assume that the government orchestrated the pattern.

In other cases, people assume that some *seed*—some preexisting, built-in inhomogeneity in the environment—gave rise to the pattern, much as a grain of sand gives rise to a pearl. When Minsky saw the pattern of aggregation in the slime-mold program, for example, he assumed that the placement of food determined the pattern.

In other words, people tend to assume that patterns are created either *by lead* or *by seed*. These tendencies were apparent in many StarLogo projects. When people observed patterns in StarLogo programs, they often assumed the existence of a leader or a seed. And when they tried to construct patterns in their own StarLogo programs, they often treated one of the turtles as the leader, or they treated some of the patches as seeds. Three brief case studies follow, showing how high-school students relied on by-lead-or-by-seed thinking as they worked on StarLogo projects.

Dead Ants

Two high-school students, Frank and Ramesh, decided to work on a StarLogo project involving ants. They weren't so much interested in how ants gather their food. Rather, they were interested in how ants gather their dead colleagues into "ant cemeteries." They had seen an ant colony where all of the dead ants were gathered neatly into piles. They wondered, How does that happen?

This question is similar to the question that Callie and I explored in chapter 3: How do termites gather wood chips into piles? The ant-cemetery problem is nearly identical, with ants taking the place of termites, and dead ants taking the place of wood chips. But Frank and Ramesh approached the problem very differently from Callie and me. Whereas Callie and I focused on decentralized strategies, Frank and Ramesh gravitated toward centralized ones.

Ramesh's first idea was that each ant could create its own cemetery, on the spot where the ant found its first dead ant. This strategy avoids all of the messiness of interaction among the ants. But at a cost: the ant colony will end up with as many cemeteries as there are live ants. In effect, Ramesh's strategy turns every ant into a leader of its own one-member colony.

Frank suggested an alternative idea: "There could be some chemical. Each turtle will look for where the chemical is. And then all turtles will put the dead ants around the patch where the chemical is." This is the *by seed* approach. One patch, with chemical, acts as the seed from which a cemetery will grow. I pointed out a problem with this approach.

Mitchel: *So there would be a few designated places with chemical. Someone or something has to decide on those places. Who is going to make that decision?*

Frank: *Maybe God? [Laughs]*

With his laugh, Frank indicated that he was aware of the problem. He knew that he could not rely on God to create the seed for the ant cemetery. But who else (or what else) could create the seed for the cemetery? After a while, Frank shifted to another idea: "Maybe the leader of this group of ants tells them where to put the dead ants." With this idea, control would still be centralized, but a leader would be in control instead of God. (Frank did not say, "Maybe the group needs a leader to decide where . . . " Rather, he just assumed that the group *must* have a leader. The only question was what the leader should do.)

Ramesh also liked the idea of a single predesignated cemetery. When I suggested that the ants might not need a predesignated cemetery, Ramesh rejected the idea: "Why would an ant try to gather other dead ants if there is no reason for it? If you make just one place, then the ants have a goal to put dead ants in one place." So, according to Ramesh, a predesignated cemetery is needed to give the ants a *goal,* a reason for collecting dead ants. In Ramesh's world view, there is no place for unintended patterns, arising from decentralized interactions. For an ordered cemetery to form, according to Ramesh, the ants must view the creation of the cemetery as an explicit goal.

Frank and Ramesh wrote several StarLogo programs based on the fixed-cemetery idea. They developed a nice strategy to help live ants find dead ants more quickly: each dead ant emitted a chemical scent, and each live ant (when not carrying a dead ant) followed the gradient of the scent. In some versions, each live ant "knew" exactly where the cemetery was. After picking up a dead ant, the ant would head directly toward the cemetery (using the StarLogo `toward` primitive). In other versions, the ant would wander randomly until it bumped into the cemetery.

After a while, I encouraged Frank and Ramesh to consider strategies without a fixed, preexisting cemetery. For a while, they considered a strategy similar to the one Callie and I tried with the termites: live ants should pick up dead ants and put them down near other dead ants. But Frank and Ramesh were very concerned about ants removing dead ants from already existing piles.

Ramesh: Once the ants place them [the dead ants], we have to set a rule that they don't get picked up again. . . . You could have one ant trying to create a pile, and another ant trying to destroy it. . . . You can't break up a pile. It has to keep getting bigger.

Callie and I had similar worries before trying out the strategy in the termite project. But Frank and Ramesh carried their concerns to an extreme, insisting on what might be called a "monotonic imperative":

piles must always grow and never shrink. They were reluctant to even try the strategy. They were sure that it was doomed to failure.

So Frank suggested a new idea. Perhaps the area with the most dead ants should automatically become the cemetery. He proposed that the screen be divided into squares: "Then the square with the most dead ants, it becomes a cemetery." This is a promising idea. In effect, a random fluctuation in the distribution of dead ants could form the seed for a cemetery. But Frank and Ramesh weren't sure how to follow through on this idea. Ramesh thought that a centralized decision maker would still be needed: "You would need an observer to decide where the most dead ants are, to make a cemetery." Again, they instinctively believed that a leader was needed to make the decision.

Ramesh suggested a new strategy. He explained, "When a certain number of [live] ants get near one another, all drop [the dead ants] at once. Once you pick up a dead ant, you try to look for other ants who have done the same thing, and try to move toward each other. If you get ten close together, they all drop their ants and start a cemetery." I asked how the live ants would find each other. Frank, remembering the slime-mold aggregation program that I had shown, suggested that the live ants use a chemical to attract one another. And how would the live ants know when to drop the dead ants? Ramesh suggested that the live ants should drop their dead ants when the chemical rose above a certain threshold, indicating lots of other ants-carrying-ants in the vicinity.

Again, this general idea seems promising. But Frank and Ramesh quickly reverted back to more centralized approaches. Ramesh suggested that the first cemetery created by the ants should become *the* cemetery. And after creating *the* cemetery, the ants should revert back to the original program (in which they pick up dead ants and take them to *the* cemetery). Frank was insistent that dead ants should never be removed from *the* cemetery: "Surround the cemetery with a fence or something, and say don't take any more dead ants from here."

"There Was Nothing There . . . "

After Ari and Fadhil created their traffic-flow simulation, I showed the simulation to other high-school students. Each car in the simulation followed three simple rules:

- If there is a car close ahead of you, slow down.

- If there aren't any cars close ahead of you, speed up (unless you are already moving at the speed limit).

- If you detect a radar trap, slow down.

All of the students quickly recognized that a traffic jam would form behind the radar trap. I asked the students what else could cause a traffic jam. They had no trouble thinking up possible causes for traffic jams: an overturned truck, a broken bridge, an entry ramp with merging traffic, a patch of ice on the road. Then I removed the radar trap from the simulation and asked what would happen. In general, the students expected the cars to end up evenly spaced along the highway, separated by equal distances. Several of them talked about the cars reaching an "equilibrium," characterized by equal spacing. No one expected a traffic jam to form. Some of their predictions:

Emily: [The cars will] just speed along, just keep going along . . . they will end up staggered, in intervals.

Frank: Nothing will be wrong with it. Cars will just go. . . . There's no obstacles. The cars will just keep going, and that's it.

Ramesh: They will probably adjust themselves to a uniform distance from each other.

When I ran the simulation, and traffic jams began to form, the students were clearly surprised. Some of them even questioned the validity of the simulation. Ramesh complained that the simulation was a "perfect world," unlike the real world.

In their comments, most students revealed a strong commitment to the idea that some type of seed (like an accident or a broken bridge) is needed to start a traffic jam. Perhaps Frank expressed it best: "I didn't think there would be any problem, since there was nothing there." In other words, if there is nothing there, if there is no seed, there shouldn't be a traffic jam. Traffic jams don't just happen; they must have localizable causes. And the cause must come from outside the system (not from the cars themselves). Some researchers who study systems talk about *exogenous* (external) and *endogenous* (internal) factors affecting the behavior of a system. In the minds of many, it seems, patterns (such as traffic jams) can be formed only by exogenous factors.

Robots and Gold

As a probe into the centralized mindset, I asked some of the high-school students to consider the following hypothetical situation (inspired by Steels 1990):

Suppose that we discover large deposits of gold on some distant planet. It is too dangerous and costly to send human astronauts to this planet, so we decide to send a spaceship with several thousand small robots. Each robot has a sensor to detect when it gets near gold, and a scoop to dig for (and carry) the gold. Once the spaceship lands on the planet, we want the robots to explore for gold and bring the gold back to the spaceship. How should we program each of the robots? In other words, what type of rules and strategies should the robots follow?

I posed this problem *after* the students had seen the StarLogo ant-foraging simulation. The two situations are very similar: the robots must collect the gold and bring it to a central location, just as the ants must collect food and bring it to the nest. The students had already seen (in the StarLogo ant program) a decentralized solution to this problem, based on ants laying pheromone trails. An analogous solution for the robots-and-gold problem might involve a collection of simple, identical robots that communicate with one another by leaving markers (such as bread crumbs) in the environment.

In presenting the robots-searching-for-gold problem, I asked the students to suggest general rules and strategies for the robots. (They did not write actual computer programs for the robots.) The students developed a variety of interesting and creative strategies. But there were certain consistencies in the responses. And in almost all cases, the students' strategies conflicted with the ants' strategies. Whereas ants use local communications, the students' robots typically had global communications capabilities. Each robot could communicate with every other robot, no matter where the robots were located, no matter how many robots were trying to communicate at the same time. Many of the robots had walkie-talkies and other high-tech forms of communication. As one student explained, "Communications without high-tech stuff is really difficult." And whereas ants rely on relative positioning, the students' robots typically had perfect knowledge of their locations on the planet. One student suggested that each robot go to a particular longitude and latitude. Once a robot found gold, it communicated its exact location to the other robots (so that they could come and help).

Most strikingly, the students' strategies were almost always centralized, relying on a leader to make decisions. Fadhil centralized control at the spaceship: "If a robot finds gold, it sends a signal to the spaceship. Then, the spaceship sends signals back to the other robots, telling them where to go. The spaceship would be constantly monitoring all of the robots." Benjamin suggested that "the leader robot should send the others in all directions, like the spokes of a wheel." Ramesh had a similar idea: "One robot is in charge, sending all these robots out. Where most gold is found, it sends more in that direction. And where the gold is not found, you eliminate that direction. So you zero in where the gold is and you get it. Cancel all the angles where there's no gold. Limit your search."

These strategies were not necessarily *wrong*; in fact, many of the student strategies seemed like they would work, and some were very creative. But it is interesting that the students were so committed to centralized approaches, even in a situation inspired by the decentralized strategies of ants, and even after the students had seen a decentralized solution to the problem.

Why the Centralized Mindset?

Why is it that people have such a strong commitment to centralized approaches? There are undoubtedly many reasons. For one thing, many phenomena in the world *are,* in fact, organized by a central designer. These phenomena act to reinforce the centralized mindset. When people see neat rows of corn in a field, they assume (correctly) that the corn was planted by a farmer. When people watch a ballet, they assume (correctly) that the movements of the dancers were planned by a choreographer. When people see a watch, they assume (correctly) that it was designed by a watchmaker.

Moreover, most people participate in social systems (such as families and school classrooms) where power and authority are very centralized (often excessively so, for my tastes). These hierarchical systems serve as strong models. Many people are probably unaware that other types of organization are even possible. In an earlier research project (Resnick 1990) I developed a programming language (called MultiLogo) based on "agents" that communicated with one another. In using the language, children invariably put one of the agents in charge of the others. One student explicitly referred to the agent in charge as "the teacher." Another referred to it as "the mother."

Perhaps most important, our intuitions about systems in the world are deeply influenced by our conceptions of ourselves. According to modern cognitive theories, our minds are composed of thousands of interacting entities, but we experience ourselves as singular selves. This is a very convenient, perhaps necessary, illusion for surviving in the world. When I do something, whether I'm painting a picture or organizing a party, I feel as if "I" am playing the role of the "central actor." It feels like there is one entity in charge: me. So it is quite natural that I should expect most systems to involve a central actor, or a single entity that is in charge.

Our images of ourselves shape what we see and what we build. Each of us experiences the world in a sequential, centralized way, so is it any surprise that early computer designers chose sequential, centralized architectures for their machines? James Bailey (1992) argues that early computers were designed in the image of the "human computers" that preceded them. He writes, "In effect, the architects of the 1940's packaged their wonderfully speedy electronic circuits in anthropomorphic forms to meet an existing market."

There was a self-reinforcing spiral. People saw the world in centralized ways, so they constructed centralized tools and models, which further encouraged a centralized view of the world. Until recently, there was little pressure against this centralization spiral. For many things that people created and organized, centralized approaches tended to be adequate, even superior to decentralized ones. Even if someone wanted to experiment with decentralized approaches, there were few tools or opportunities to do so.

But the centralization spiral is now starting to unwind. As organizations and scientific models grow more complex, there is a greater need for decentralized ideas. And new decentralized tools (like StarLogo) are emerging that enable people to actually implement and explore such ideas. Thus the stage is set to move beyond the centralized mindset.

Beyond the Centralized Mindset

The centralized mindset is deeply entrenched. When people see patterns and structures, they instinctively assume centralized causes or centralized control. They often see leaders and seeds where none exist. When something happens, they assume that one individual agent must be responsible.

But the centralized mindset is neither unchanging nor unchangeable. As decentralized ideas infiltrate the culture—through new technologies, new organizational structures, new scientific ideas—people will undoubtedly begin to think in new ways. People will become familiar with new models and new metaphors of decentralization. They will begin to see the world through new eyes. They will gradually recraft and expand their ways of thinking about causality. At times, they will still appeal to the traditional centralized explanations. But when those explanations don't work, they will have other models and metaphors to draw on.

My work with StarLogo provides initial glimpses of how people can begin to move beyond the centralized mindset. In this section I discuss how people, as they played with StarLogo, became engaged (both intellectually and emotionally) with decentralized phenomena—and with new types of thinking associated with decentralized phenomena.

The Allure of Decentralization

As I watched people working with StarLogo, I became aware of a seeming contradiction. On one hand, StarLogo users tended to assume centralized causes and control, assuming that StarLogo patterns must be formed by lead or by seed. Especially when they first began using StarLogo, people had difficulty writing decentralized StarLogo programs, or even recognizing decentralized phenomena as such.

On the other hand, most StarLogo users were fascinated—almost mesmerized—with self-organizing phenomena when they observed them on the screen. These phenomena seemed to have a strong emotional pull. People seemed drawn to them, even if they misunderstood them. There was an apparent tension: people felt a gut attraction to decentralized phenomena, even as they clung tightly to centralized preconceptions. For some, this emotional engagement with decentralized phenomena acted as a foundation, a starting point, for moving beyond the centralized mindset.

The gut attraction to decentralized phenomena can be seen in the wild popularity of "the wave" at sporting arenas. The wave is formed by spectators themselves, as they stand up and sit down at the appropriate times. Everyone participates. People stand up at their seats when the wave reaches them, then sit down as it sweeps past. There is no conductor or choreographer for the wave. No one is in charge. The wave is a rare opportunity for people to create and participate in a self-organizing phenomena. And they are clearly excited by it. The wave was first seen at sporting arenas just a decade ago, but it is now a mainstay at all types of athletic competitions, from high-school through professional.

Part of the attraction of the wave is that you get a lot for a little. Each individual does nothing more than stand up and sit down (at the appropriate times), but together the actions produce a giant wave. Many StarLogo users had the same sort of feeling about StarLogo programs. One user said that he felt like he was "cheating." It didn't seem fair to get so much for so little. Frank described it this way: "In this version of Logo, you can get more than what you tell it to do." Benjamin had a similar reaction: "It's pretty simple. That's what I like about this. It's weird. You can build a simple little program, and the things that it does, what you can do with it. I mean there's not much to that program. If you were working with regular Logo, to do something even halfway. . . . I mean, there are so many procedures and stuff you have to build with regular Logo. Here, with a few short procedures, a lot happens. I don't know how to put it exactly."

At one point Benjamin worked on a termite program, in which a leader termite told the other termites what to do. After a while I showed Benjamin the leaderless termite program that Callie and I had written. His reaction was, "This seems simpler. Everything happens *automatically*." For Benjamin, the decentralized approach seemed almost magical, getting something for (almost) nothing.

The unpredictable nature of decentralized StarLogo programs had a strong appeal for some users. Of course, unpredictability is not unique to decentralized or parallel programs. Traditional Logo programs can do some pretty unexpected things. But StarLogo programs often involve a particularly intriguing type of unpredictability: users can understand fully what each individual object will do, but have no sense of what the overall system will do.

In general, those who thrived in the StarLogo environment were those who relished (not resisted) unpredictability. Callie is an example. At one point, while we were struggling to get our termite program working, I asked Callie if we should give up on our decentralized approach and program the termites to take their wood chips to predesignated spots. Callie quickly dismissed this suggestion:

Mitchel: *We could write the program so that the termites know where the piles are. As soon as a termite picks up a wood chip, it could just go to the pile and put it down.*

Callie: *Oh, that's boring!*

Mitchel: *Why do you think that's boring?*

Callie: *Cause you're telling them what to do.*

Mitchel: *Is this more like the way it would be in the real world?*

Callie: *Yeah. You would almost know what to expect if you tell them to go to a particular spot and put it down. You know that there will be three piles. Whereas here, you don't know how many mounds there are going to be. Or if the number of mounds will increase or decrease. Or things like that. . . . This way, they [the termites] made the piles by themselves. It wasn't like they [the piles] were artificially put in.*

For Callie, preprogrammed behavior, even if effective, was "boring." Callie preferred the decentralized approach since it made the termites seem more independent ("they made the piles by themselves") and less predictable ("you don't know how many mounds there are going to be").

Sherry Turkle (1984) writes of the computer's "holding power." That holding power seems particularly strong when people are playing with (or even just watching) decentralized phenomena on the computer. Many people were transfixed by the StarLogo slime-mold program, in which turtles organize themselves into clusters. The turtles were represented by simple dots of light, not actual turtle images. That left lots of room for interpretation and imagination. For different people, the simulation evoked different images.

When I showed the program to an economist, he was reminded of the development of cities.

When I showed the program to an educational researcher, she talked about interactions among children in a classroom. She discussed how students can organize themselves, and form their own learning communities, without a dominating teacher.

When I showed the program to a student at the Sloan School of Management at MIT, it reminded her of information flowing through an organization. She talked about the advantages and disadvantages of decentralized information systems.

When I showed the program to a Zen student, he saw the turtles as people in search of religion. He was intrigued that the turtles formed smaller groups when they sniffed in more directions. "When people become more perceptive, they don't have to rely on big groups anymore," he noted. "People join big groups to satisfy needs that they can't satisfy on their own."

When people constructed their own StarLogo programs, the holding power seemed even stronger. Many users related to the action on the screen in very personal ways. Two graduate students worked with StarLogo for several sessions as part of a project for a class at Harvard. They created a simple ecosystem with a desert and a jungle. Turtles were more likely to die when they were in the desert, and more likely to reproduce when they were in the jungle. The two students stared at the screen for nearly an hour as the turtles scampered about. They cheered as the population increased, groaned as the population shrank. When the population made a brief spurt, they were careful not to let their expectations rise too much. "It's toying with us," one of them warned. The students compared themselves to the relatives of a hospital patient, watching the heart monitor alongside the patient's bed. When the turtles finally went extinct, it was as if the heart monitor stopped beeping. "They can't ever come back," said one of the "grieving" students.

Emily, a high-school student, worked on a similar project. She divided her StarLogo world into nine regions, each with a different climate. Each region had its own temperature and level of rainfall. Turtles reproduced depending on how "happy" they were with the local climate. Like the graduate students, Emily seemed deeply invested in the fates of the turtles. She saw her StarLogo program as very different from other computer programs she had used. She explained, "It's always doing something different. You can come back a week later, and something different could be going on. You can stand here and look at it for a long time. With other computer programs, you walk in and look at it, and for the first five minutes you're fascinated, then you wander off. And later you come back and look at it, and you think 'Uh, that again.'"

Guiding Heuristics for Decentralized Thinking

Although they were clearly intrigued with decentralized phenomena, many StarLogo users struggled to understand what they saw on the screen. Through those struggles, certain ideas emerged as very useful in making sense of decentralized phenomena. These ideas came up again and again, in many different situations. They served as "guiding heuristics" for thinking about decentralized worlds.

In this section I discuss five of these heuristics:

• *Positive Feedback Isn't Always Negative.* Positive feedback often plays an important role in creating and extending patterns and structures.

• *Randomness Can Help Create Order.* Most people view randomness as destructive, but in some cases it actually helps make systems more orderly.

• *A Flock Isn't a Big Bird.* It is important not to confuse levels. Often, people confuse the behaviors of individuals and the behaviors of groups.

• *A Traffic Jam Isn't Just a Collection of Cars.* It is important to realize that some objects ("emergent objects") have an ever-changing composition.

• *The Hills Are Alive.* People often focus on the behaviors of individual objects, overlooking the environment that surrounds the objects.

These guiding heuristics are not very "strong." They are neither prescriptive nor predictive, nor are they unique to decentralized systems. They don't tell you precisely how to think about decentralized systems, nor do they tell you how to make accurate predictions about such systems. Rather, they are ideas to keep in mind as you try to make sense of an unfamiliar system, or to design a new one. They highlight some pitfalls to avoid, and some possibilities not to overlook. These heuristics will not (and should not) replace the centralized mindset. Rather, they can supplement the centralized mindset, leading to a richer set of mental models for making sense of systems in the world.

Positive Feedback Isn't Always Negative

Positive feedback has an image problem. People tend to see positive feedback as destructive, making things spiral out of control. Positive feedback is symbolized by the awful screeching sound that results when a microphone is placed near a speaker. By contrast, negative feedback is viewed as very useful, keeping things under control. Negative feedback is symbolized by the thermostat, which keeps room temperature at a desired level by turning the heater on and off as needed.

Historically, researchers have paid much more attention to negative feedback than to positive feedback. As Jean-Louis Deneubourg and Simon Goss (1989) note, "When feedback is discussed in animal groups, it is nearly always negative feedback that is considered, and its role is limited to that of a regulatory mechanism, in which fluctuations are damped and equilibrium is the goal. . . . Positive feedback is only rarely considered." Brian Arthur (1990) notes a similar bias in economic research.

The negative image of positive feedback is even part of popular culture. A few years ago, I heard Johnny Carson tell what might be called a positive feedback joke. Carson referred to a scientific study about the depletion of the Earth's ozone layer. He noted that the depletion was caused, in part, by use of chlorofluorocarbons, typically found in spray cans containing, for example, antiperspirants. Carson said he had come up with a theory: People used antiperspirants, which caused depletion of the ozone layer, which caused temperatures on Earth to go up, which caused people to use more antiperspirant, which caused further depletion of the ozone layer, which caused further increases in temperature, which . . .

When I asked high-school students about positive feedback, most weren't familiar with the term. But they were certainly familiar with the concept. When I explained what I meant by positive feedback, the students quickly generated examples. Not surprisingly, almost all of their examples involved something getting out of control, often with destructive consequences. One student talked about scratching a mosquito bite, which made the bite itch even more, so she scratched it some more, which made it itch even more, and so on. Another student talked about stock market crashes: a few people start selling, which makes more people start selling, which makes even more people start selling, and so on.

Despite these negative images, positive feedback often plays a crucial role in self-organizing phenomena. Economist Brian Arthur (1990) points to the geographic distribution of cities and industries as an example of a self-organizing process driven by positive feedback. As soon as a small nucleus of high-technology electronics companies started in Santa Clara County south of San Francisco, an infrastructure developed to serve the needs of those companies. That infrastructure encouraged even more electronics companies to locate in Santa Clara County, which encouraged the development of an even more robust infrastructure. And thus Silicon Valley was born.

Many StarLogo explorations rely on similar positive feedback mechanisms for creating patterns and structures. In the slime-mold program, a few cells wander near one another and form a small pheromone puddle, which attracts more cells, which drop even more pheromone, which makes the puddle bigger, which attracts even more cells, and so on. A similar mechanism is involved in the ant-foraging program: a few ants discover some food and form a faint pheromone trail to the nest, which attracts even more ants, which reinforce the trail on their way back to the nest, and so on.

Of course, negative feedback is involved in these programs too. In the slime-mold program there is a limited supply of slime-mold cells. As the clusters grow larger, there are fewer free cells to join the clusters. This puts a negative feedback control on the growth of the clusters. But it is positive feedback that creates and extends the structures in the first place.

For some students who used StarLogo, the idea of positive feedback provided a new way of looking at their world. One day, Fadhil came to me excitedly. He had been in downtown Boston at lunchtime, and he had a vision. He imagined two people walking into a deli to buy lunch. "Once they get their food, they don't eat it there. They bring it back with them. Other people on the street smell the sandwiches and see the deli bag, and they say, 'Hey, maybe I'll go to the deli for lunch today!' They were just walking down the street, minding their own business, and all of the sudden they want to go to the deli. As more people go to the deli, there's even more smell and more bags. So more people go to the deli. But then the deli runs out of food. There's no more smell on the street from the sandwiches. So no one else goes to the deli."

Randomness Can Help Create Order
Like positive feedback, randomness has a bad image. Most people see randomness as annoying at best, destructive at worst. They view randomness

in opposition to order: randomness undoes order, it makes things disorderly.

This view of randomness was apparent in some reactions to StarLogo projects. I showed the StarLogo turtle-geometry projects to several of the high-school students. In one case I put several thousand turtles at the same position but gave all of the turtles random headings. I asked the students what would happen if all of the turtles moved forward 50 steps. One student responded, "Each turtle has a random heading, so they'll go all over the place." In his mind, randomness was clearly associated with disorder ("all over the place"). Even after seeing the turtles move outward in an expanding circle, one of the students remained bothered: "If the turtles have random headings, why are they always forming a circle?"

Despite its image as "antiorder," randomness plays an important role in many self-organizing systems. People often assume that seeds are needed to initiate patterns and structures. When people see a traffic jam, for example, they assume the traffic jam grew from a seed—perhaps a broken bridge or a radar trap. In general, this is a useful intuition. The problem is that most people have too narrow a conception of seeds. They think only of preexisting inhomogeneities in the environment—like a broken bridge on the highway, or a piece of food in an ant's world.

This narrow view of seeds causes misintuitions when people try to make sense of self-organizing systems. In self-organizing systems, seeds are neither preexisting nor externally imposed. Rather, self-organizing systems often create *their own* seeds. It is here that randomness plays a crucial role. In many self-organizing systems, random fluctuations act as the seeds from which patterns and structures grow.

In the StarLogo traffic program, no traffic jams form if the cars are given equal initial velocities and spaced evenly along the highway. But if there is some randomness in either the initial velocities or positions, small density fluctuations (that is, fluctuations in the density of cars) will develop along the highway: a few more cars along one stretch of the road, a few fewer cars along another stretch. These density fluctuations serve as the seeds for traffic jams. Positive feedback accentuates these density fluctuations, making the seeds sprout into full-fledged traffic jams.

The situation is similar in other StarLogo projects. In the segregation project random fluctuations in the densities of turtles and frogs are reinforced by positive feedback, producing single-species clusters. In the slime-mold project the seeds of clusters are formed when a few slime-mold cells happen to wander near one another, causing a random fluctuation in the density of pheromone. Then positive feedback takes over.

Regions with higher densities of pheromone attract more slime-mold cells, causing the pheromone density to rise still higher, attracting even more slime-mold cells, and so on.

This combination of random fluctuations plus positive feedback underlies many everyday phenomena. Sometimes, at concerts or sporting events, thousands of spectators join together in rhythmic, synchronized clapping. There is no conductor leading them. How do they coordinate their applause? Here's one way to think about what happens. Initially, when everyone starts clapping, the applause is totally unorganized. Even people clapping at the same tempo are wildly out of phase with one another. But, through some random fluctuation, a small subset of people happen to clap at the same tempo, in phase with one another. That rhythm stands out, just a little, in the clapping noise. People in the audience sense this emerging rhythm and adjust their own clapping to join it. Thus the emerging rhythm becomes a little stronger, and even more people conform to it. Eventually, nearly everyone in the audience is clapping in a synchronized rhythm. Amazingly, the whole process takes just a few seconds, even with thousands of people participating.

Randomness plays yet another role in some self-organizing processes—it makes possible the exploration of multiple options. Ant researcher Jean-Louis Deneubourg notes that ants do not follow pheromone trails perfectly. Instead, ants have a probabilistic chance of losing their way as they follow the trails. Deneubourg and his colleagues (1986) argue that this "ant randomness" is *not* a defective stage on an evolutionary path "towards an idealistic deterministic system of communication." Rather, this randomness is an evolutionarily adaptive behavior. Deneubourg describes an experiment with two food sources near an ant nest: a rich food source far from the nest, and an inferior source close to the nest. Initially, the ants discover the inferior food source and form a robust trail to that source. But some ants wander off the trail. These "lost ants" discover the richer source and form a trail to it. Since an ant's pheromone emissions are related to the richness of the food source, the trail to the richer source becomes stronger than the original trail. Eventually, most ants shift to the richer source. So the randomness of the ants provides a way for the colony to explore multiple food sources in parallel. While positive feedback encourages *exploitation* of particular sources, randomness encourages *exploration* of multiple sources.

The StarLogo ant program exhibits a somewhat similar phenomenon. While most of the ants are exploiting one food source, some "lost ants" often discover (and form a weak trail to) a more distant food source. Most of the colony's ants continue to exploit the closer food source. (Unlike the Deneubourg experiment, the more distant food source is not "richer" in any sense.) But the weak trail to the more distant food source serves a useful purpose. When the closer food source is fully depleted (and its associated pheromone trail evaporated), the ants that had been exploiting that source are free to look for other food. The weak trail formed by the lost ants acts as a seed for a new trail, and it is quickly reinforced by the newly freed ants. So when the colony finishes with one food source, it doesn't have to start from scratch to find a new one. The randomness of the ants allows the colony to continue to explore all of the time, even as most of the ants are exploiting the food source that is currently most attractive.

The StarLogo slime-mold project provides another example of randomness in the service of exploration. If the program had no randomness, slime-mold cells would rarely leave their clusters. The program would lose its dynamic and organic quality. The screen would become filled with lots of little clusters, with little or no interchange of cells between clusters. The randomness in the program makes it more likely for cells to break free of their clusters. As a result, small clusters become less stable: when a small cluster loses one of its cells, the whole cluster is likely to break apart. Small clusters either grow or break apart. The result is fewer, larger clusters, with more cells moving from cluster to cluster.

If the goal is for the slime-mold cells to aggregate into large clusters (as is the case with real slime mold), then randomness plays a very useful role. The situation with lots of small clusters can be seen as a "local optimum" for the slime-mold system: each cell is happy (since each cell is in a cluster), but the overall system is not (since the clusters are too small). In effect, the random motion of the slime-mold cells ensures that the system doesn't get stuck on such a local optimum. Instead, the randomness induces the system to explore for a more "global" optimum (larger clusters).

A Flock Isn't a Big Bird

In trying to make sense of decentralized systems and self-organizing phenomena, the idea of *levels* is critically important. Interactions among objects at one level give rise to new types of objects at another level. Interactions among slime-mold cells give rise to slime-mold clusters. Interactions among ants give rise to foraging trails. Interactions among cars give rise to traffic jams. Interactions among birds give rise to flocks.

In many cases, the objects on one level behave very differently from objects on another level. In the StarLogo traffic program, for example, traffic jams tend to move backward, even though all of the cars within the jams are moving forward. Ari, one of the two students who wrote the StarLogo traffic program, was not very surprised by the backward motion of the traffic jam. He made an analogy to the StarLogo ant-colony program which he had seen earlier. "It's sort of like the ants," he explained. "They get together as one body. All sorts of little ones get together and form a big thing. So each of the cars is forming a huge mass, like a blob, which can move either backward or forward regardless of how the cars are moving."

Ari clearly distinguished between levels: he expected ant colonies to act differently from individual ants, traffic jams to act differently from individual cars. But other high-school students, upon seeing the StarLogo traffic program, found the backward motion of the traffic jams surprising, or at least strange. As Emily put it, "When you try to visualize it, it seems kind of strange. But when you see it, it looks logical." Ramesh had a stronger reaction. He insisted that "real" traffic jams wouldn't move backward. He argued that the backward motion of the jam must be an artifact of the way the cars "wrapped" around the edges of the computer screen. In real traffic jams, he argued, "the car leaving from the front of the jam eliminates the possibility of the jam moving back. The jam goes with the cars."

Frank, Ramesh's partner, had a different problem. He expected the traffic jams to exhibit some type of simple periodic motion. Why? Because the program controlling the individual cars consisted of simple looping constructs. Frank explained, "A loop does the same thing every time. So the whole thing should be repeating itself." Frank, like Ramesh, was confusing levels. Just because the behavior of each car is controlled by a simple loop, the behavior of the traffic jam will not necessarily "loop" in a simple way.

Confusion of levels is not a problem restricted to scientifically naive high-school students. I showed the StarLogo traffic program to two visiting researchers, each of whom is involved in the cybernetics research community. They were not at all surprised that the traffic jams were moving backward. They were well aware of that phenomenon. But then one of the researchers said, "You know, I've heard that's why there are so many accidents on the freeways in Los Angeles. The traffic jams are moving backward and the cars are rushing forward, so there are lots of accidents." The other researcher thought for a moment, then replied, "Wait a minute. Cars crash into other cars, not into traffic jams." In short, he felt that the first researcher had confused levels, mixing cars and jams inappropriately. The two researchers then spent half an hour trying to sort out the problem.

Traffic jams are hardly a special case. People confuse levels in many situations. Consider the StarLogo program with turtles and frogs sharing a pond. Each individual creature is relatively tolerant, so it seems natural to expect an even mix of turtles and frogs throughout the pond. But as with cars and traffic jams, the group behavior in the pond is very different from individual behaviors. The overall population ends up much more segregated that any of the individuals really wants.

Or consider what happens in the growth of a plant. Focus first on the cellular level. Cells on the dark side of the plant produce more of the hormone auxin than cells on the light side. Auxin promotes cell elongation. So the dark side of the plant grows more quickly than the side facing the light. It is tempting to say that the plant "likes" the dark. But now think about the plant as a whole. Since the dark side of the plant grows more quickly, the whole plant bends toward the light. (Imagine a vertical bar in which the right side expands faster than the left side. By the geometry of the situation, the bar must bend toward the left.) So at this higher level, it is tempting to say that the plant "likes" the light. Of course, neither conclusion is right or wrong. It all depends on which level you focus on.

A Traffic Jam Isn't Just a Collection of Cars

For most everyday objects, it is fair to think of the object as a collection of particular parts (a particular chair might have four particular legs, a particular seat, a particular back). But not so with objects like traffic jams. Thinking of a traffic jam as a collection of particular parts is a sure path to confusion. The cars composing a traffic jam are always changing, as some cars leave the front of the jam and other join from behind. Even when all of the cars in the jam are replaced with new cars, it is still the same traffic jam.

Many objects in StarLogo programs have this same quality. Just as the cars in a traffic jam are always changing, so too are the cells in a slime-mold cluster. Objects like traffic jams and slime-mold clusters can be thought of as "emergent objects"—they emerge from the interactions among lower-level objects (such as cars and slime-mold cells). It is the interactions among the lower-level objects, not the particular lower-level objects themselves, that define emergent objects. We can even think of ourselves as emergent objects: within my body, old cells are always dying and new cells are being created, but I remain the same person.

Frank and Ramesh's difficulties with their ant-cemetery project were due, in large part, to their difficulties in thinking about emergent objects. They were adamant that dead ants should never be taken from a cemetery because they thought the dead ants defined the cemetery. How can a cemetery grow, they wondered, if the dead ants in it are continually being taken away? In fact, if Frank and Ramesh had relaxed their "monotonic imperative" and allowed the composition of ant cemeteries to vary with time (as Callie and I allowed the composition of the wood-chip piles to vary in the termite project), they probably would have been much more successful in their project.

The issue of emergent objects came up in another form in the same ant-cemetery project. At one point, Frank and Ramesh wanted each live ant to be surrounded by a chemical scent (so that other ants could detect it). One way to do that is to make the ants release a continuous stream of chemical into the environment (that is, onto the StarLogo patches), and to make the chemical diffuse and evaporate with time. No matter how or where an ant moves, it will always be surrounded by a fresh "halo" of chemical. Each ant's halo is an emergent object: its composition is always changing, as the chemical evaporates and the ant releases new chemical.

But Frank and Ramesh didn't think to do it that way. They tried to make each ant "carry" a nondiffusing halo of chemical with it, as if the chemical were part of its clothing. In their approach the halos were *not* emergent objects: the composition of each halo never changed. As it turned out, their approach was much more difficult. If they had thought of emergent halos, their task would have been much easier.

The Hills Are Alive

In *Sciences of the Artificial* (1969), Herbert Simon describes a scene in which an ant is walking on a beach. Simon notes that the ant's path might be quite complex. But the complexity of the path, says Simon, is not necessarily a reflection of the complexity of the ant. Rather, it might reflect the complexity of the beach.

Simon's point: don't underestimate the role of the environment in influencing and constraining behavior. People often seem to think of the environment as something to be *acted upon,* not something to be *interacted with.* People tend to focus on the behaviors of individual objects, ignoring the environment that surrounds (and interacts with) the objects.

A richer view of the environment is particularly important in thinking about decentralized and self-organizing systems. So in designing StarLogo, I explicitly tried to highlight the environment. By introducing patches as a new class of object, I hoped to encourage people to view the environment in new ways. In traditional versions of Logo the world is like a passive piece of paper, just waiting for turtles to draw on it. In StarLogo the world is alive—it can execute actions even as turtles move on top of it. The StarLogo world is full of interactions: interactions between turtles and turtles, between patches and patches, between turtles and patches. Through all of these interactions, large-scale patterns can arise.

The environment serves different roles in different StarLogo programs. In the turtle-ecology program the patches are responsible for growing new food. In the forest-fire program the patches are responsible for spreading the fire from tree to tree. In several programs the patches are responsible for making chemicals evaporate and diffuse.

StarLogo's active environment is particularly useful as a type of communications medium. Patches can spread messages to other turtles and patches. In the ant-foraging program the patches spread two types of chemical messages. One chemical acts as an indirect communication between ants—one ant drops the chemical, other ants sense it. A second chemical spreads outward from the nest, making it possible for ants to find their way home. Many people think that ants must need sophisticated memories, or some type of nonlocal communication with the nest, in order to find their way home. But by using the environment effectively, ants can find the nest with simple rules and local interactions.

Some students were quick to exploit this new form of communication. When Ari and Fadhil started working on the traffic-jam project, StarLogo had only "nearest neighbor" interactions between turtles and patches. So cars could "see" only one car-length ahead. To allow cars to see several car-lengths ahead, Ari and Fadhil worked out the following approach: each car emitted some "exhaust" to the patch directly behind it, then the patches spread the exhaust backward (patch-to-patch) the desired number of times. Then if a car detected exhaust in the patch directly ahead of it, it meant that there must be another car close ahead. (Ari and Fadhil changed their program after I added the `patch-polar` primitive to StarLogo. That primitive allows turtles to sense nonlocally, sniffing an arbitrary number of patches away in any arbitrary direction. But Ari and Fadhil's original car-exhaust approach seems more elegant, since it fits the local-interaction spirit of StarLogo.)

On the other hand, some students resisted the idea of an active environment. When I explained the StarLogo ant-foraging program to Frank, he was worried that pheromone trails would continue to attract ants even after the food sources at the ends of the trails had been fully depleted. He developed an elaborate scheme in which the ants, after collecting all of the food, deposited a second pheromone to neutralize the first pheromone. It never occurred to him to let the first pheromone evaporate away. In his mind the ants had to take some positive action to get rid of the first pheromone. They couldn't rely on the environment to make it go away.

Looking Ahead: From Reflections to Projections

There is an apparent paradox in people's reactions to decentralized systems. On one hand is the allure of decentralization. People are intrigued and inspired by decentralized systems. They are fascinated by systems that are organized without an organizer, coordinated without a coordinator.

On the other hand is the centralized mindset. When people see patterns in the world, they intuitively assume that the patterns are created either by lead or by seed. And when they try to design patterns, they start with the same assumptions.

What will happen in the future? As decentralized ideas spread through the culture, will people continue to cling tightly to their centralized assumptions? Next, in Projections, I speculate about the future.

5

Projections

The centre cannot hold.

—W. B. Yeats, *The Second Coming*

Growing Up

A friend of mine has a daughter named Rachel. By the time Rachel was three years old, she had already developed a theory about why it rains on some days and not on others. She explained, "The clouds rain when the thunder *commands* them to rain" (emphasis hers). In Rachel's mind, some type of centralized decision making was necessary. Thunder commanded, and the clouds obeyed.

It is not surprising that Rachel came up with a centralized explanation for the rain. Most likely, she was unaware that other types of explanations even exist. But what will happen as Rachel grows up? Will she continue to rely on centralized explanations? If she takes a physics course in high school, will she understand gravity as two objects pulling on one another with equal force? Or will she think of gravity as a one-way force, with one large object pulling on a smaller one? If she takes an economics course in college, will she understand that interest rates and the money supply can each affect one another? Or will she assume that one is the cause and the other is the effect? As an adult, will she believe that creationism is the only reasonable explanation for the origin of the species? If the unemployment rate goes up, will she immediately assume some type of evil conspiracy? Or will she search for explanations with multiple, interacting causes?

What will influence Rachel's thinking on these issues? If she takes a new high-school course on decentralized thinking, in which she is taught ten Golden Rules of decentralized thinking, would that make much of a difference? Maybe some, but probably not much. Being taught a list of rules isn't going to have much effect on a firmly entrenched centralized mindset. Rachel is likely to move beyond the centralized mindset only if she participates in a culture that values and encourages decentralized thinking. One isolated high-school course isn't enough. Decentralized ideas must spread to all school courses—and to life outside of school.

New computational tools can play an important role in the spread of decentralized ideas. Rachel is likely to become comfortable with decentralized ideas only if she gets opportunities to design, create, explore, and play with decentralized systems. There are already some commercial software packages that allow you to play with decentralized phenomena. With one product, called SimCity, you can experiment with urban development. Put a housing development here, build a police station there, adjust the tax rate, and see if the city prospers. Another product, called SimAnt, allows you to experiment with ant-colony behavior. Dig a new tunnel here, drop a little pheromone there, adjust the ratio of worker ants to breeder ants, and see if the colony survives.

These software packages are a start, but they are often constraining. You can't change the underlying models that control the simulations, nor can you change the underlying context. What if you are interested in neither urban development nor ant behavior, but in cars and traffic? Today you are out of luck. What's needed are *microworld construction kits,* so that you can create your own microworlds, focusing on the domains you find most interesting. StarLogo is a step in that direction; more sophisticated microworld construction kits are sure to follow.

How will Rachel use microworld construction kits? At school, she might create artificial ecosystems with giraffes and elephants, her favorite animals. At home, she and her friends might create a simulation of how people gather into groups at a party. By working on projects like these, Rachel will feel an increasing sense of ownership over decentralized ideas. Gradually, she will become comfortable with new ways of thinking.

Rachel will not necessarily abandon all of her centralized strategies—nor should she. In economics, a unyielding commitment to decentralized, laissez-faire strategies can be just as debilitating as an unyielding commitment to centralized planning. So too with thinking: an unyielding decentralized mindset is no better than a centralized one. Many phenomena in the world *do* have centralized explanations. Many phenomena *are* caused by lead or by seed. New decentralized ideas should supplement, not supplant, centralized strategies. As Rachel constructs theories about the world, she should be able to draw on both centralized and decentralized ideas.

By the time Rachel was four, she had developed a new theory about the rain. "The clouds get together at night, and they decide whether or not it should rain the next day," she explained. This new theory still involves some centralized planning. But there is no longer a central actor (the thunder) in charge of the whole process. As Rachel continues to grow, surrounded by new types of computational tools and new types of ideas, one can only wonder what new theories she'll develop to explain the rain.

Appendix A
Student Participants

About one dozen high-school students participated in the "explorations" described in this book. To recruit students, I visited classrooms, gave a brief overview of StarLogo, and asked for volunteers. I recruited students at one local high school (Woburn High School), and at several summer-school and weekend courses for high-school students (at MIT and Radcliffe).

The students had widely varying levels of experience with computers, and they came from a variety of ethnic backgrounds—roughly half were immigrants or first-generation Americans. About two-thirds of the students were male, one-third female. One thing that all of the participants shared was a willingness to come to MIT to work on an experimental project. All student names used in this book are pseudonyms.

Students typically came to MIT for eight to ten sessions, each lasting 60 to 90 minutes. Most students worked together in pairs. I worked directly with the students, suggesting projects, asking questions, challenging assumptions, helping with programming, and encouraging students to reflect on their experiences as they worked with StarLogo. Computer interactions were saved in computer files, and all discussions were recorded on audiotape.

In the early sessions I typically showed students existing StarLogo programs. The students experimented with the programs, trying different parameters and making slight modifications of the programs. As the sessions progressed, I encouraged students to develop their own projects, based on personal interests.

I provided each student with a collection of magazine and journal articles on topics related to decentralized systems and self-organizing phenomena. Ideas from these articles often served as the basis for student projects. The following articles were included:

Arthur, W. B. (1990). "Positive Feedbacks in the Economy," *Scientific American*, vol. 262, no. 2 (February), pp. 92–99.

Dewdney, A. K. (1989). "Simulated Evolution: Wherein Bugs Learn to Hunt Bacteria," *Scientific American*, vol. 260, no. 5 (May), pp. 138–141.

Franks, N. (1989). "Army Ants: A Collective Intelligence," *American Scientist*, vol. 77 (March–April), pp. 139-145.

Herman, R., and Gardels, K. (1963). "Vehicular Traffic Flow," *Scientific American*, vol. 209, no. 6 (December), pp. 35–43.

Seeley, T. (1989). "The Honey Bee Colony as a Superorganism," *American Scientist*, vol. 77 (November–December), pp. 546-553.

Appendix B

StarLogo Overview

This appendix provides a brief overview of StarLogo, describing the major features of the language. It is not intended as a full definition of the StarLogo language. Currently, there are versions of StarLogo for the Connection Machine, UNIX workstations, and Macintosh computers. Details of the language vary from one implementation to another. So if you want to run the programs presented in this book, you will probably need to make some modifications to the programs. Consult the manual for the particular version of StarLogo that you are using.

The Cast of Characters

StarLogo includes three main types of "characters" (or, in computer-science parlance, three classes of objects):

• *Turtles.* Following the Logo tradition, "turtles" are the main inhabitants of the StarLogo world. But StarLogo turtles go beyond traditional Logo turtles in several ways: you can control thousands of turtles; all of the turtles can execute commands at the same time (in parallel); there are new built-in procedures to control interactions among the turtles (and between the turtles and their "world"); turtles can "clone" new turtles; you can give different traits to different turtles (using "state variables").

• *Patches.* Patches are "pieces" of the world in which the turtles live. Patches are not merely passive objects upon which the turtles act. Like turtles, patches can execute StarLogo commands. Patches are arranged in a grid—similar to cellular automata. So StarLogo is somewhat like a cellular-automata world with turtles roaming around on top.

• *Observer.* The observer "looks down" on the turtles and patches. The observer can create new turtles, and it can monitor the activity of turtles and patches.

Sample Commands

(Note: Many StarLogo commands have abbreviated forms. For example, you can use `setc` instead of `set-color`, and `fd` instead of `forward`. In this book I generally use the full command names, to make the code more readable.)

`create-turtle 100`
100 turtles appear on the screen. By default, the turtles start with random positions and random headings.

`forward 200`
All turtles move forward 200 "turtle steps." The turtles "wrap" at the edges of the screen. The turtles do not draw as they move: unlike traditional Logo turtles, StarLogo turtles start with their "pens" up.

`forward random 200`
Each turtle chooses a different random number (between 0 and 199), so each moves forward a different distance.

`if ypos < 0 [set-color green]`
The position (0,0) is at the center of the screen. So only turtles in the bottom half of the screen have y-positions (ypos) less than 0. Those turtles turn green.

`set-patchcolor yellow`
Each patch sets its color to yellow. So the whole "background" of the screen turns yellow.

`if patch-xpos > 20 [set-patchcolor green]`
Each patch with an x-position greater than 20 turns green.

`if (distance 10 20) < 15 [set-patchcolor white]`
Each patch checks to see if its distance from the point (10,20) is less than 15 units. If so, it turns white. The result: a white disk of radius 15, centered on the point (10, 20).

Communications

The StarLogo procedure `ask` is used for communicating between StarLogo objects (turtles, patches, and observer). An object can `ask` another object (turtle or patch or observer) to perform a particular action, or it can `ask` another object for some information. (Note: `ask` has a different meaning in traditional versions of Logo.) The procedure `ask` takes two inputs: the first input indicates the recipient of the communication (who), and the second input indicates the content of the communication (what).

There are many ways to indicate the recipient of the communication. You can explicitly supply the ID number of the turtle or patch. (The ID number is returned by the primitive procedure who.) But in most cases, you will want to use an "ID-generating procedure." For example, patch-here generates the ID of the patch underneath the turtle, and turtle-here generates the ID of a turtle within a given patch. patch 0 generates the ID of the patch directly to the north, and turtle 0 generates the ID of the turtle directly to the north (if one exists). Note that patch 0 and turtle 0 can be used by either turtles or patches.

Other ID-generating procedures include turtle-at and patch-at (which take absolute xy-coordinates as inputs), patch-xy (which takes relative xy-coordinates as inputs), and patch-polar (which takes relative polar coordinates as inputs).

Sample Commands

```
if (ask patch-here [chemical > 5]) [set-color white]
```
Each turtle asks the patch underneath it if its value for chemical is greater than 5; if so, the turtle turns white. (The parentheses in the expression are optional; they are included to make the expression easier to read.)

```
ask patch heading + 180 [set-chemical chemical + 1]
```
Each turtle tells the patch directly behind it to increase its value for pheromone by 1.

Demons

Demons are short programs that continuously run in the "background." By using StarLogo demons, you can make many different things happen at the same time, and you can interact with StarLogo programs while they are running. You can create as many demon programs as you would like, and they all execute continuously and (roughly) simultaneously. In this way, you can simulate a (simple) type of "process parallelism" to go along with the "data parallelism" of StarLogo.

The execution of demons is based on the StarLogo "clock." This clock keeps "ticking" whenever StarLogo is not executing a top-level command. At every tick of the clock, StarLogo automatically runs each demon procedure.

You can selectively turn on and off particular demons for particular turtles or patches, using the StarLogo commands `activate-demon` and `deactivate-demon`.

Breeds

In some situations, it is useful to have several "breeds" of creatures (in addition to turtles). For example, you might want foxes and rabbits in an ecology model, or antibodies and antigens in an immunology model.

You create new breeds with the `create-breed` command. For example, to create a new breed called `ant`, you execute `create-breed ant`. When you execute this command, StarLogo automatically generates a procedure named `create-ant` for creating ants. Each new breed is automatically assigned a different default color. Creatures of the new breed can use all of the standard StarLogo procedures for turtles (such as `fd` and `rt` and `uphill`).

StarLogo automatically generates a set of new procedures for interacting with creatures of the new breed. For example, it generates the procedures `ant-total`, `ant?`, and `ant-here`. Here is general rule: for each standard StarLogo procedure that includes the word `turtle`, StarLogo will create a new procedure with the name of the new breed. (StarLogo also has procedures with the word `creature` in the place of `turtle`; these procedures refer to all creatures, regardless of breed. For example, `creature-total` reports the total number of creatures, regardless of breed.)

Colors

StarLogo has 256 colors. These colors are organized into 25 "major" colors, with ten intensities (or shades) of each. This organization is useful for achieving smooth shading effects (using the `scale-color` command).

The `set-color` command uses the following number scheme for the colors: 0 for black, 1–10 for red, 11–20 for green, 21–30 for blue, 31–40 for yellow, and so on. For each major color, the highest number is the "brightest" (closest to white); the lowest number is the "darkest" (closest to black). For example, 20 is the brightest green; 11 is the darkest green.

To see part of the color table, type `set-patchcolor patch-ypos`. The screen will fill with horizontal lines of color, a different color for each *y*-position.

References

Abelson, H., and diSessa, A. (1980). *Turtle Geometry: The Computer as a Medium for Exploring Mathematics*. Cambridge, MA: MIT Press.

Abelson, H., and Goldenberg, P. (1977). "Teacher's Guide for Computational Methods of Animal Behavior." Logo Memo no. 46. MIT Artificial Intelligence Laboratory, Cambridge, MA.

Abelson, H., and Sussman, G. (1985). *Structure and Interpretation of Computer Programs*. Cambridge, MA: MIT Press.

Adamson, E., and Helgoe, C. (1989). "Exploring Art and Technology." *Proceedings of the EuroLogo Conference*, Gent, Belgium. Amsterdam: IOS, 1990.

Agre, P. (1991). *The Dynamic Structure of Everyday Life*. Cambridge: Cambridge University Press.

Ambrose, S. (1992). "Writers on the Grassy Knoll: A Reader's Guide." *New York Times Book Review*, February 2, p. 1.

Arthur, W. B. (1990). "Positive Feedbacks in the Economy." *Scientific American*, vol. 262, no. 2 (February), pp. 92–99.

Bailey, J. (1992). "First We Reshape Our Computers, Then Our Computers Reshape Us: The Broader Intellectual Impact of Parallelism." *Daedalus*, vol. 121, no. 1 (Winter), pp. 67–86.

Barlow, J. P. (1992). "The Great Work." *Communications of the ACM*, vol. 35, no. 1 (January), pp. 25–27.

Bateson, G. (1972). *Steps to an Ecology of Mind*. New York: Ballantine Books.

Belenky, M., Clinchy, B., Goldberger, N., and Tarule, J. (1986). *Women's Ways of Knowing: The Development of Self, Voice, and Mind*. New York: Basic Books.

Braitenberg, V. (1984). *Vehicles*. Cambridge, MA: MIT Press.

Brinch Hansen, P. (1975). "The Programming Language Concurrent Pascal." *IEEE Trans. Software Eng.*, vol. 1, no. 2, pp. 199–207.

Brooks, R. (1991). "Intelligence without Representation." *Artificial Intelligence*, vol. 47, pp. 139–160.

Cohen, J., and Kelly, F. (1990). "A Paradox of Congestion in a Queuing Network." *Journal of Applied Probability*, vol. 27, no. 3, pp. 730–734.

Cohen, M., and Hagan, P. (1981). "Diffusion-Induced Morphogenesis in *Dictyostelium*." *Journal of Theoretical Biology*, vol. 93, pp. 881–908.

Collins, R., and Jefferson, D. (1991). "AntFarm: Towards Simulated Evolution." In *Artificial Life II*, edited by C. Langton, C. Taylor, J. D. Farmer, and S. Rasmussen. Redwood City, CA: Addison-Wesley.

Darwin, C. (1859). *On the Origin of Species by Means of Natural Selection or the Preservation of Favored Races in the Struggle for Life* (often referred to as *The Origin of Species*). London: Murray.

Dawkins, R. (1986). *The Blind Watchmaker*. New York: W. W. Norton.

Deneubourg, J. L., Aron, S., Goss, S., Pasteels, J. M., and Duerinck, G. (1986). "Random Behavior, Amplification Processes, and Number of Participants: How They Contribute to the Foraging Properties of Ants." *Physica D*, vol. 22, pp. 176–186.

Deneubourg, J. L., and Goss, S. (1989). "Collective Patterns and Decision-Making." *Ethology, Ecology, & Evolution*, vol. 1, pp. 295–311.

Deneubourg, J. L., Pasteels, J. M., and Verhaeghe, J. C. (1983). "Probabilistic Behavior in Ants: A Strategy of Errors?" *Journal of Theoretical Biology*, vol. 105, pp. 259–271.

Dennett, D. (1991). *Consciousness Explained*. Boston: Little, Brown, and Co.

DeParle, J. (1991). "For Some Blacks, Social Ills Seem to Follow White Plans." *New York Times*, August 11, Week in Review, p. 5.

Dewdney, A.K. (1989). "Simulated Evolution: Wherein Bugs Learn to Hunt Bacteria." *Scientific American*, vol. 260, no. 5 (May), pp. 138–141.

Drescher, G. (1987). "Object-Oriented Logo." In *Artificial Intelligence and Education*, edited by R. Lawler and M. Yazdani. Norwood, NJ: Ablex Publishing.

Dumaine, B. (1990). "Who Needs a Boss?" *Fortune*, May 7, pp. 52–60.

Dumaine, B. (1991). "The Bureaucracy Busters." *Fortune*, June 17, pp. 36–47.

Dumanoski, D. (1991). "Challenging the Big Myth of Our Past." *The Boston Globe*, January 14, p. 25.

Evans, R. (1991). StarLogo-D: Process Parallelism and Programming Environments Through Demons. Undergraduate thesis, Department of Electrical Engineering and Computer Science, MIT, Cambridge, MA.

Farmer, D., and Packard, N. (1986). "Evolution, Games, and Learning: Models for Adaptations in Machines and Nature." *Physica D*, vol. 22D, no. 1.

Fikes, R., and Nilsson, N. (1971). "STRIPS: A New Approach to the Application of Theorem Proving to Problem Solving." *Artificial Intelligence*, vol. 2, pp. 189–208.

Forrest, S., ed. (1991). *Emergent Computation*. Cambridge, MA: MIT Press.

Forrester, J. (1971). *World Dynamics*. Cambridge, MA: Wright-Allen Press.

Franks, N. (1989). "Army Ants: A Collective Intelligence." *American Scientist*, vol. 77 (March–April), pp. 139–145.

Fukuyama, F. (1992). *The End of History and the Last Man*. New York: MacMillan.

Gardner, M. (1970). "The Fantastic Combinations of John Conway's New Solitaire Game 'Life'." *Scientific American*, vol. 223, no. 4 (April), pp. 120–123.

Gilligan, C. (1982). *In a Different Voice: Psychological Theory and Women's Development*. Cambridge, MA: Harvard University Press.

Gleick, J. (1987). *Chaos: Making a New Science*. New York: Viking Penguin Inc.

Goldberg, A., and Robson, D. (1983). *Smalltalk-80: The Language and Its Implementation*. Reading, MA: Addison-Wesley.

Goldbeter, A., and Segel, L. A. (1977). "Unified Mechanism for Relay and Oscillation of Cyclic AMP in Dictyostelium Discoideum." *Proc. of the National Academy of Sciences*, vol. 74, pp. 1543–1547.

Goss, S., Beckers, R., Deneubourg, J. L., Aron, S., and Pasteels, J. M. (1990). "How Trail Laying and Trail Following Can Solve Foraging Problems for Ant Colonies." In *Behavioral Mechanisms of Food Selection*, edited by R. N. Hughes. Berlin: Springer-Verlag.

Granott, N. (1990). "Puzzled Minds and Weird Creatures: Spontaneous Inquiry and Phases in Knowledge Construction." In *Constructionism*, edited by I. Harel and S. Papert. Norwood, NJ: Ablex Publishing.

Halstead, R. (1985). "Multilisp: A Language for Concurrent Symbolic Computation." *ACM Trans. of Prog. Languages and Systems*, vol. 7, no. 4, pp. 501–538.

Harel, I. (1991). *Children Designers*. Norwood, NJ: Ablex Publishing.

Harel, I., and Papert, S. (1990). "Software Design as a Learning Environment." *Interactive Learning Environments*, vol. 1, no. 1, pp. 1–32.

Harvey, B. (1985). *Computer Science Logo Style*. Cambridge, MA: MIT Press.

Heppner, F., and Grenander, U. (1990). "A Stochastic Nonlinear Model for Coordinated Bird Flocks." In *The Ubiquity of Chaos*, edited by S. Krasner. Washington, DC: AAAS Publications.

Herman, R., and Gardels, K. (1963). "Vehicular Traffic Flow." *Scientific American*, vol. 209, no. 6 (December), pp. 35–43.

Hillis, W. D. (1985). *The Connection Machine*. Cambridge, MA: MIT Press.

Hofstadter, D. (1979). *Gödel, Escher, Bach: The Eternal Golden Braid*. New York: Basic Books.

Hogeweg, P. (1989). "MIRROR beyond MIRROR, Puddles of LIFE." In *Artificial Life*, edited by C. Langton. Redwood City, CA: Addison-Wesley.

Holldobler, B., and Wilson, E. O. (1990). *The Ants*. Cambridge, MA: Harvard University Press.

Huberman, B., ed. (1988). *The Ecology of Computation*. New York: Elsevier.

Hut, P., and Sussman, G. J. (1987). "Advanced Computing for Science." *Scientific American*, vol. 255, no. 10.

Keller, E. F. (1985). *Reflections on Gender and Science*. New Haven, CT: Yale University Press.

Keller, E. F., and Segel, L. (1970). "Initiation of Slime Mold Aggregation Viewed as an Instability." *Journal of Theoretical Biology*, vol. 26, pp. 399–415.

Kernan, A. (1990). *The Death of Literature*. New Haven: Yale University Press.

Kohlberg, L. (1981). *The Philosophy of Moral Development*. San Francisco: Harper and Row.

Kuhn, T. (1962). *The Structure of Scientific Revolutions*. Chicago: University of Chicago Press.

Langton, C., ed. (1989). *Artificial Life*. Redwood City, CA: Addison-Wesley.

Lave, J. (1988). *Cognition in Practice: Mind, Mathematics, and Culture in Everyday Life*. Cambridge: Cambridge University Press.

Leiser, D. (1983). "Children's Conceptions of Economics—The Constitution of a Cognitive Domain." *Journal of Economic Psychology*, vol. 4, pp. 297–317.

Lindgren, K. (1991). "Evolutionary Phenomena in Simple Dynamics." In *Artificial Life II*, edited by C. Langton, C. Taylor, J. D. Farmer, and S. Rasmussen. Redwood City, CA: Addison-Wesley.

Lotka, A. J. (1925). *Elements of Physical Biology*. New York: Dover Publications (reprinted 1956).

Lovelock, J. (1979). *Gaia: A New Look at Life on Earth*. New York: Oxford University Press.

Malone, T., Yates, J., and Benjamin, R. (1987). "Electronic Markets and Electronic Hierarchies." *Communications of the ACM*, vol. 30, no. 6 (June), pp. 484–497.

Martin, F. (1988). "Children, Cybernetics, and Programmable Turtles." Master's Thesis, MIT Media Laboratory, Cambridge, MA.

Mayr, E. (1982). *The Growth of Biological Thought*. Cambridge, MA: Harvard University Press.

McKenna, R. (1985). *The Regis Touch*. Reading, MA: Addison-Wesley.

Miller, R. (1991). "Still on the Farm, 200 Million Ants Later." *New York Times*, Dec. 19, p. C2.

Minsky, M. (1987). *The Society of Mind*. New York: Simon & Schuster.

Minsky, M., and Papert, S. (1988). Epilogue from *Perceptrons* (Expanded Edition). Cambridge, MA: MIT Press. (Original edition published 1969.)

Nicolis, G., and Prigogine, I. (1989). *Exploring Complexity*. New York: W. H. Freeman and Company.

Paley, W. (1802). *Natural Theology—or Evidences of the Existence and Attributes of the Deity Collected from the Appearances of Nature*. Oxford: J. Vincent.

Palmiter, M. (1989). *Simulated Evolution*. Bayport, NY: Life Science Associates.

Pancake, C. (1991). "Software Support for Parallel Computing: Where Are We Headed?" *Communications of the ACM*, vol. 34, no. 11 (November), pp. 53–64.

Papert, S. (1980). *Mindstorms: Children, Computers, and Powerful Ideas*. New York: Basic Books.

Papert, S. (1991a). "Situating Constructionism." In *Constructionism*, edited by I. Harel and S. Papert. Norwood, NJ: Ablex Publishing.

Papert, S. (1991b). "Perestroika and Epistemological Politics." In *Constructionism*, edited by I. Harel and S. Papert. Norwood, NJ: Ablex Publishing.

Peterson, I. (1988). *The Mathematical Tourist: Snapshots of Modern Mathematics*. New York: W. H. Freeman.

Prigogine, I., and Stengers, I. (1984). *Order out of Chaos: Man's New Dialogue with Nature*. New York: Bantam Books.

Resnick, M. (1989). "LEGO, Logo, and Life." In *Artificial Life*, edited by C. Langton. Redwood City, CA: Addison-Wesley.

Resnick, M. (1990). "MultiLogo: A Study of Children and Concurrent Programming." *Interactive Learning Environments*, vol. 1, no. 3, pp. 153–170.

Resnick, M. (1991). "Xylophones, Hamsters, and Fireworks: The Role of Diversity in Constructionist Activities." In *Constructionism*, edited by I. Harel and S. Papert. Norwood, NJ: Ablex Publishing. (Also available as Epistemology and Learning Memo no. 9, MIT Media Laboratory, Cambridge, MA.)

Resnick, M. (1992). "StarLogo Manual." Epistemology and Learning Group, MIT Media Laboratory, Cambridge, MA.

Resnick, M., and Martin, F. (1991). "Children and Artificial Life." In *Constructionism*, edited by I. Harel and S. Papert. Norwood, NJ: Ablex Publishing. (Also available as Epistemology and Learning Memo no. 10, MIT Media Laboratory, Cambridge, MA.)

Resnick, M., and Ocko, S. (1991). "LEGO/Logo: Learning through and about Design." In *Constructionism*, edited by I. Harel and S. Papert. Norwood, NJ: Ablex Publishing. (Also available as Epistemology and Learning Memo no. 8, MIT Media Laboratory, Cambridge, MA.)

Resnick, M., Ocko, S., and Papert, S. (1988). "LEGO, Logo, and Design." *Children's Environments Quarterly*, vol. 5, no. 4.

Resnick, M., and Sargent, R. (1992). UNIX StarLogo (software). Epistemology and Learning Group, MIT Media Laboratory, Cambridge, MA.

Reynolds, C. (1987). "Flocks, Herds, and Schools: A Distributed Behavioral Model." *Computer Graphics*, vol. 21, no. 4, pp. 25–36.

Roberts, N., Anderson, D., Deal, R., Garet, M., and Shaffer, W. (1983). *Introduction to Computer Simulation: A System Dynamics Modeling Approach.* Reading, MA: Addison-Wesley.

Robson, S., and Resnick, M. (1991). "The Use of Massively Parallel Computers to Investigate Emergent Properties and Group-Level Phenomena in Social Insects." Presentation at the annual meeting of the Entomological Society of America. Reno, NV, December.

Rogoff, B., and Lave, J., eds. (1984). *Everyday Cognition: Its Development in Social Context.* Cambridge, MA: Harvard University Press.

Rorty, R. (1979). *Philosophy and the Mirror of Nature.* Princeton, NJ: Princeton University Press.

Rumelhart, D., McClelland, J., and the PDP Research Group. (1986). *Parallel Distributed Processing.* Cambridge, MA: MIT Press.

Sabot, G. (1988). *The Paralation Model.* Cambridge, MA: MIT Press.

Salomon, G. (1992). *Distributed Cognitions: Psychological and Educational Considerations.* Cambridge: Cambridge University Press.

Schelling, T. (1971). "On the Ecology of Micromotives." *The Public Interest*, no. 25, pp. 61–98.

Seeley, T. (1989). "The Honey Bee Colony as a Superorganism." *American Scientist*, vol. 77 (November–December), pp. 546–553.

Segal, H. (1978). *Introduction to the Work of Melanie Klein.* London: Hogarth Press.

Semler, R. (1989). "Managing without Managers." *Harvard Business Review*, September–October, pp. 76–84.

Senge, P. (1990). *The Fifth Discipline: The Art & Practice of the Learning Organization.* New York: Doubleday/Currency.

Silverman, B. (1987). *Phantom Fishtank.* Montreal: Logo Computer Systems Inc.

Simon, H. (1969). *The Sciences of the Artificial.* Cambridge, MA: MIT Press.

Smith, A. (1776). *An Inquiry into the Nature and Causes of the Wealth of Nations* (often referred to as *The Wealth of Nations*). Indianapolis: Hackett Pub. Co., 1993.

Steels, L. (1989). "The RDL Manual." AI Memo 89-11. University of Brussels.

Steels, L. (1990). "Cooperation between Distributed Agents through Self-Organization." In *Decentralized A.I.*, edited by Y. Demazeau and J. P. Muller. Amsterdam: North-Holland.

Tinbergen, N. (1951). *The Study of Instinct*. Oxford: Oxford University Press.

Toffoli, T., and Margolus, N. (1987). *Cellular Automata Machines*. Cambridge, MA: MIT Press.

Travers, M. (1989). "Animal Construction Kits." In *Artificial Life*, edited by C. Langton. Redwood City, CA: Addison-Wesley.

Turkle, S. (1984). *The Second Self: Computers and the Human Spirit*. New York: Basic Books.

Turkle, S. (1988). "Artificial Intelligence and Psychoanalysis: A New Alliance." *Daedalus*, vol. 117, no. 1 (Winter), pp. 241–268.

Turkle, S., and Papert, S. (1990). "Epistemological Pluralism." *Signs*, vol. 16, no. 1 (Autumn).

Varela, F., Thompson, E., and Rosch, E. (1991). *The Embodied Mind: Cognitive Science and Human Experience*. Cambridge, MA: MIT Press.

Verity, J. (1991). "Out of One Big Blue, Many Little Blues." *Business Week*, December 9, p. 33.

Volterra, V. (1926). "Fluctuations in the Abundance of a Species Considered Mathematically." *Nature,* vol. 188, pp. 558–560.

von Bertalanffy, L. (1968). *General System Theory*. New York: George Braziller.

von Foerster, H., Mead, M., and Teuber, H., eds. (1949). *Cybernetics: Circular, Causal, and Feedback Mechanisms in Biological and Social Systems: Transactions of the Eighth Conference*. New York: Josiah Macy, Jr. Foundation, 1952.

Walter, W. G. (1950). "An Imitation of Life." *Scientific American*, vol. 182, no. 5 (May), pp. 42–45.

Wiener, N. (1948). *Cybernetics, or Control and Communication in the Animal and the Machine*. New York: John Wiley.

Wilensky, U. (1991). "Abstract Meditations on the Concrete." In *Constructionism*, edited by I. Harel and S. Papert. Norwood, NJ: Ablex Publishing. (Also available as Epistemology and Learning Memo no. 12, MIT Media Laboratory, Cambridge, MA.)

Wilson, E. O. (1971). *The Insect Societies*. Cambridge, MA: Harvard University Press.